# SpringerBriefs in Psychology

## SpringerBriefs in Cognition

*Series editor*

Gesine Dreisbach, Fakultät für Psychologie, Universität Regensburg, Regensburg, Germany

SpringerBriefs in Cognition is a subseries of SpringerBriefs in Psychology aimed to publish a wide range of books that present comprehensive and up-to-date overviews of current developments in narrow topics pertaining basic and applied cognitive domains. The series covers low-level sensory, attentive, perceptual and motor processes, and high-level cognitive functions including learning, memory, language, communication, decision, thinking, cognitive control and action planning. Contributions from the fields of psychology, neuroscience and cognitive science are welcome in this series, provided that they deepen our understanding on cognition and behavior in basic or applied contexts. Specific theoretical or methodological topics, including systematic evaluation of alternative theories and state of the art on emerging theories are expected. Integrative research reviews that develop connections between different areas of research are particularly valuable. Both solicited and unsolicited proposals are considered for publication in this series. All proposals will be subject to peer review by external referees.

Series Editor Bio:
Dr. Gesine Dreisbach is Chair for General and Applied Psychology at the University of Regensburg. Her research focuses on processes of cognitive control, with a particular focus on task switching and task rules; context-sensitive adjustment of cognitive control; conflicts as aversive signals for control adaptations; and affective and motivational modulation of cognitive control.

For detailed Instructions for Authors, please email series editor Prof. Dr. Gesine Dreisbach at gesine.dreisbach@psychologie.uni-regensburg.de

More information about this series at http://www.springer.com/series/10737

Máximo Trench • Ricardo A. Minervino

# Distant Connections: The Memory Basis of Creative Analogy

Springer

Máximo Trench
Psychology Department
IPEHCS, CONICET- Universidad
del Comahue
Bariloche, Rio Negro, Argentina

Ricardo A. Minervino
National Scientific and Technical
Research Council
IPEHCS, CONICET-Universidad
del Comahue
Cipolletti, Rio Negro, Argentina

ISSN 2192-8363 ISSN 2192-8371 (electronic)
SpringerBriefs in Psychology
ISSN 2625-2929 ISSN 2625-2937 (electronic)
SpringerBriefs in Cognition
ISBN 978-3-030-52547-7 ISBN 978-3-030-52545-3 (eBook)
https://doi.org/10.1007/978-3-030-52545-3



This Springer imprint is published by the registered company Springer Nature Switzerland AG
The registered company address is: Gewerbestrasse 11, 6330 Cham, Switzerland

*From Máximo*
*To my parents, who always supported my dubious endeavors, and to my cheerful daughter Lucía, from whom I borrowed the time to complete this work*
*From Ricardo*
*To Bruno and Lorenzo, my lovely sons, for having the good grace of not demanding me the time that my insensate dedication to work has stolen from them*

# Acknowledgments

Our sincere gratitude to Pedro Minervino for patiently satisfying almost tyrannical restrictions in crafting our capricious illustrations. We are especially grateful to Mark Keane, Ruth Byrne, Adam Green, Brian Ross, Kenneth Kurtz, Valeria Olguín, and Micaela Tavernini for generous feedback and encouragement and to Robert Goldstone, Douglas Hofstadter, and David Landy for opening their labs to us at Indiana University, Bloomington. We are also indebted to Dedre Gentner and Keith Holyoak for laying the foundations of this exciting line of inquiry.

# Contents

# About the Authors

**Máximo Trench** is professor of Psychology at the National University of Comahue and adjunct researcher at the National Council for Scientific and Technical Research. He has taught graduate-level courses of cognitive processes at several Masters and Ph.D. programs (Universidad de Buenos Aires, Universidad Nacional del Sur, Universidad de Palermo, FLACSO, and UCES) and has served as visiting scholar at Indiana University Bloomington with support from the Fulbright Commission and the Argentine Ministry of Education. His research is conducted both at the Bariloche campus of the University of Comahue and at the Patagonic Institute for Studies in the Humanities and the Social Sciences. His interests concentrate on the mechanisms underlying the generation and interpretation of analogical comparisons, as they take place during educationally relevant activities such as learning, problem-solving, argumentation, hypothesis generation, and comprehension. His intra- and extramural projects aim at developing cognitive interventions to increase access to analogically related learning that would otherwise remain inert in long-term memory.

**Ricardo A. Minervino** is professor of Cognitive Psychology at the National University of Comahue and independent researcher at the National Council for Scientific and Technical Research of Argentina. He has taught graduate-level courses of thinking and creativity in several Masters and Ph.D. programs (Universidad Nacional del Comahue, Universidad de Buenos Aires, Universidad Nacional del Sur, Universidad Nacional de la Plata, and FLACSO). His research is conducted both at the School of Education of the University of Comahue and at the Patagonic Institute for Studies in the Humanities and the Social Sciences. His current projects aim at determining the mechanisms that are involved in the production of a particular kind of analogies—those that are framed by schema-governed categories—as well as the role that different pragmatics play in these processes. His broader interests extend to the role of conceptual metaphors in cognition and to the role of analogies in creativity.

# Chapter 1
# Introduction

Taking advantage of numerous planetary observations collected in Tycho Brahe's observatory over decades, his young successor Johannes Kepler noticed that the period of a planet's translation increased with its distance from the Sun. But Kepler's notions of motion were still Aristotelian: as a perfect vacuum is impossible, every moving object needs a force to propel it. Hence, assuming all planets to have similar propulsion capabilities, longer periods could simply reflect that the outer planets need to cover a longer distance. But as it turned out, farther planets not only displayed lower angular speeds, but lower *tangential* speed: they simply *moved slower*. How could their intelligences be coordinated in such a manner that farther planets adjusted their self-propulsion according to their distance from the Sun? Avid to replace this kind of teleological explanations by the type of causation that characterized modern Physics, Kepler discarded self-regulated propulsion in favor of a more parsimonious account, wherein the movement of the planets originates in a single rotational force (vis motrix) emanated from the Sun. But how did his planetary observations fit within this new scheme? To make sense of the observed regularities, he drew on his (by those days) advanced account of how light propagated in space: Just like the density of light, which gets thinner and thinner the farther the object is separated from the light source, so the motive force emanated from the Sun diminishes as planets become increasingly distant, hence propelling them to a lesser degree. While Kepler's causal explanation is incorrect by Newtonian standards, his light-vis motrix analogy helped precipitating an unparalleled shift in the history of astronomy: that from ascribing free-will to celestial entities to embracing a purely physical account based on soulless inanimate forces.

In the domain of biology, Charles Darwin had documented the transformation of species over time during extensive travel to Africa and South America. But it was not until much later that he devised an explanation of how such an intriguing process could actually take place. On the one hand, the way breeders combined extant exemplars of pigeons to generate better or novel breeds served as a model of how



M. Trench, R. A. Minervino, *Distant Connections: The Memory Basis of Creative Analogy*, SpringerBriefs in Psychology,
https://doi.org/10.1007/978-3-030-52545-3_1

the diversity of species might arise from random combinations of traits. But crucially, ideas posited by economist Thomas Malthus served as a source of inspiration for understanding how the emergence of diversity could possibly lead to a slow but coherent evolution over great many generations. Malthus had observed that while unchecked population growth is geometric, food resources grow arithmetically, inevitably leading to a "fight for survival." If one applies this idea to natural species at large, competition for limited resources would inevitably lead to the survival of those who are more fit to their environment, who will in turn produce offspring similarly characterized.

Distant analogies also figure prominently in other prototypical arenas of creative ideation, such as developing original products or solving pressing problems with existing products. As a notable example of the former case, the invention of the voice telegraph (i.e., the telephone) by Alexander Graham Bell drew heavily on an analogy to the anatomy of the human ear. In developing aids for the deaf, Alexander's father had already delved into the relation between the timbre of vowels and consonants and the shape of their corresponding sound waves. Based on a previous device of his own that used actual tympana and ossicles to record verbal sounds on a plaque of smoked glass (i.e., the *ear phonoautograph*), Alexander analogized to the structure of the human ear to suggest that if sound could be converted to an electrical current, then such current could be reconverted back to an acoustic signal at the other end of the electrical circuit, thus giving rise to a speaking telegraph. As a recent instance of improving existing products, the pharmaceutical industry faced the need of reducing the frequency with which pills had to be administered in a way that kept the concentration of a substance stable over longer periods of time. Dr. Robert Langer, inducted in 2006 to the National Inventors Hall of Fame, asked "What if we could develop a polymer that had a surface like soap—a surface that could erode harmlessly?" Together with his labmates at MIT, he drew on this analogy as a basis for developing a polymer which could be delivered either through a pill, an injection, or even an embedded wafer. As he notes, "Just like soap, the polymer we developed dissolved layer by layer, which makes it very, very safe to use in the body" (Gelb & Miller Caldicott, 2007, p. 97).

Besides their utility for generating creative products or ideas, analogies represents a powerful means of conveying ideas to others. President George H. W. Bush compared Saddam Hussein's invasion to defenseless Kuwait with Hitler's invasion of Poland at the onset of World War II: Just like Hitler, who was not satisfied with having annexed Poland and went on to conquer Europe at large, so would Saddam invade other countries, leading to a conflict of unforeseeable dimensions. Within educational settings, analogies are strategically employed to convey abstract or complex ideas in more familiar terms, as when the intangible nature of electricity is rendered more graspable through being compared to the circulation of water, or when the complex interrelation between cellular organelles is illustrated by comparison with the main components of an industrial factory.

But the utility of distant analogies for rendering the abstract more tangible far exceeds their educational deployment. As demonstrated by a wealth of studies arising after Lakoff and Johnson's (1980) seminal monograph, we inadvertently make sense of abstract concepts like *love*, *theory*, and *argument* by analogy to more concrete, perceptual-like domains such as *traveling*, *building*, or *war*, respectively. As an example, conventional expressions like "Your criticisms were right on target" or "we lost ground after the lawyer's allegation" reveal a subtle mapping between the domain of war relationships and that of argument, in terms of which arm marksmanship corresponds to the appropriateness of a claim, and moving back in the battlefield corresponds to the manifestation of weaknesses in our own assertions. Some of the conceptual metaphors that confer meaning to our abstract concepts date back centuries or even millennia. However, current proposals within conceptual metaphor theory claim that even when inadvertently producing or interpreting this kind of expressions, we neurally revive the original mapping between the concrete and the abstract domains, transferring core semantic content from the former to the latter in real-time. Recent research also points to the role of analogical comparisons in the genesis of relational categories. Unlike exemplars of entity categories like *fruit* or *bird*, which tend to be perceptually similar to each other, exemplars of relational categories like *assault* or *award* seldom share readily perceivable similarities. Upon analogizing across successive exemplars of a relational category, the learner begins to distill the core relational semantics of the concept, thus filtering out idiosyncratic traits.

On certain occasions, analogies help us better comprehend situations that we had not been able to appropriately categorize. As an example, a friend of the authors started watching soccer at home, at social gatherings, and even at work, in a way that ended up compromising his job and his marriage. The realization that his situation somehow resembles the behavior of an alcoholic may lead him to re-represent it as a case of *addiction*. This recategorization, in turn, may help him modify his situation via adapting strategies that help alcoholics avoid consumption, such as not hanging out with other consumers or staying away from alcohol outlets.

The activity of making analogies can also help in empathizing with another person's feelings and emotions. At about age ten, one of the sons of the second author got infuriated with him due to recurrently postponing the repair of his PlayStation. Only after recalling a personal experience wherein a technician took too long to fix his personal computer, could he appropriately understand his son's insistence with the PlayStation.

In view of the variety and prevalence of the activities that analogical thinking subserves, several authors have contended that analogical reasoning lies at the core of human cognition at large (Hofstadter & Sander, 2013; Holland, Holyoak, Nisbett, & Thagard, 1986) and creative cognition in particular (Boden, 1990; Hofstadter, 1985). But if analogical mechanisms can potentially give rise to outcomes ranging from the revolutionary to the mundane, then research on these cognitive mechanisms can bear on a heated debate about the nature of creativity: whether the mental processes underlying creative breakthroughs are inherently different from those employed by laypeople or else slightly optimized versions of the same set of heuristics (Ward, Smith, & Finke, 1999). The present monograph will delve into a critical component of creative analogizing: whether and how our memory systems support access to distant analogous situations stored in memory. But before moving into the interphase wherein memory and reasoning join forces to engender creative connections, we shall briefly review what is known about the central analogical subprocesses of *mapping* and *inference*, as they operate in the comparatively simpler scenario wherein two analogous situations are available in consciousness for being compared.

The remainder of the chapter will thus be organized as follows. After framing analogy research in the broader context of Cognitive Science, we will review the propositional formalisms adopted by cognitive scientists to convey and ultimately simulate how humans process structured representations. Taking advantage of this conventional language, we will characterize the concept of analogy in the context of related types of similarity matches. After discussing central empirical findings about how the mapping and inference stages of analogical thinking are computed, we conclude by exposing the way in which computational algorithms have simulated such behavioral findings. Having a glimpse of the computational complexity involved in interpreting seemingly straightforward comparisons such as that between Saddam Hussein and Hitler will be instrumental in dimensioning the paramount challenge of distant analogical retrieval, which will be the focus of the remaining chapters.

## 1.1 Analogy Research Within the Broader Enterprise of Cognitive Science

Modern research on analogical thinking is mainly rooted in Psychology (e.g., Dorolle, 1949; Duncker, 1926; Herbart, 1898) and Philosophy of Science (e.g., Black, 1962; Hesse, 1966). However, it was not until the end of the 1970s that it got its most decisive impulse based on the progress of Cognitive Science. A basic tenet of the Cognitive Science approach is that psychological theories should be specified to the level of computational processes and representations, be them symbolic, connectionist or mixed. This way, the approach grants a central role to computational simulation as a means for validating theories, a role no less important than the traditional empirical methods of experimental psychology. During the 1970s, Cognitive Science witnessed the development of a great number of studies focusing on the propositional format of representations (e.g., Miller & Johnson-Laird, 1976; Rumelhart & Norman, 1981; Schank, 1972; Schank & Abelson, 1977). These investigations were linked with studies on metaphor arising from Cognitive Psychology (Gentner, 1977a, 1977b; Ortony, 1979a, 1979b), as well as with studies on analogy arising from Artificial Intelligence (e.g., Carbonell, 1981; Winston, 1980). At about the same time, Sternberg (1977a, 1977b) relaunched research on proportional analogies in the A:B::C:? form, widely involved in the psychometric tradition (e.g., Spearman, 1927) and in developmental psychology (e.g., Piaget, Montangero, & Billeter, 1977). In light of the confluence of these lines of inquiry, many researchers in Cognitive Science welcomed the study of analogical thinking to the computational-representational understanding of the mind, on the grounds that the formalisms that were available to tackle it were both psychologically acceptable and computationally rich (Keane, 1988).

This favorable state of affairs laid the foundation for the development of the two most important theories in the field of analogical thinking from the 1980s to the present: the structure-mapping theory (Gentner, 1983, 1989) and the multiconstraint theory (Holyoak, 1984; Holyoak & Thagard, 1989; Hummel & Holyoak, 1997). Their computational models of mapping and inference generation (SME—structure mapping engine—Falkenhainer, Forbus, & Gentner, 1989; ACME—analogical constraint mapping engine—Holyoak & Thagard, 1989; LISA—learning and inference with schemas and analogies—Hummel & Holyoak, 1997) embodied their conceptions about what is an analogy, and which are the essential processes involved in producing and interpreting this kind of comparisons.

Analogies are part of our pervasive tendency to detect apparent or profound similarities between objects, facts, and situations (Gentner, 1989; Hofstadter & Sander, 2013; Holyoak & Thagard, 1995). Although the abovementioned theories employ different vocabularies, they share the distinction between different kinds of similarity, as well as an agreement on the position of analogy in this classification. Both theories shared the idea that analogical thinking operates on symbolic representations stored in a propositional format, and they ground their classification of different forms of similarity in some distinctions that this representational format allows.

## 1.2 Analogies Within a Broader Class of Similarity Relations

Propositions have a predicative character, that is, they state something about something. Every proposition includes a predicate and one or more arguments. For example, the meaning of the sentence "The car is beautiful" can be expressed through the following proposition:

1. Beautiful (car)

In this proposition the predicate *beautiful* (what is stated) is applied to *car* (the thing to which the statement applied). The predicate takes only one argument, a slot in the proposition that is filled by *car*. In the following proposition this role is filled by another vehicle:

2. Nice (motorcycle)

Entities are individual objects or beings. Predicates that take only one argument are called *attributes* since they typically describe properties of entities. If someone makes a comparison between the car and the motorcycle by virtue of their appearance, she would be carrying out an *attribute match*.

Relations are predicates that take two or more arguments, and they describe events or states. The propositional format allows decomposing relations into their roles, permitting a conceptual distinction between roles and their fillers, as well as the binding of fillers to roles. This potentiality makes it almost unavoidable for the formulation of psychological theories and computational models of analogical thinking (Doumas & Hummel, 2005). The following propositions express relations between entities:

3. Boast (Peter, motorcycle, friend)
4. Brag (Susan, car, cousin)

A comparison between the facts described in 3 and 4 constitutes a *relational match*. While first-order relations take entities as arguments, higher-order relations take other relations as arguments, giving place to systems of relations. The following are examples of this last type of structures:

5. Cause [Boast (Peter, motorcycle, friend), take-without-permission (friend, motorcycle, Peter)]
6. Cause [Brag (Susan, car, cousin), Steal (cousin, car, Susan)]

We have finally arrived at the concept of *analogy*: analogy is a comparison of two situations that share a similar system of relations (Gentner, 1983, 1989; Holyoak & Thagard, 1989, 1995). On top of maintaining similar systems of relations, analogies could vary in terms of the extent to which corresponding entities maintain shared attributes. While a comparison between the episodes described in 5 and 6 involves similar entities (e.g., the car and the motorcycle are vehicles and are both nice), another analogous situation might involve dissimilar ones (e.g., the case of a scientist whose novel ideas got plagiarized as a result of having bragged about them). In the analogical literature, this distinction has been referred to in terms of *distant* vs. *near* analogies, *superficially similar* vs. *superficially dissimilar* analogies, or *intra-domain* vs. *interdomain* analogies. Structure-mapping theory restricts the concept of analogy to comparisons in which corresponding entities do not share object properties and applies the concept of *literal simile* to analogous situations maintaining object similarities.

Mapping is the central process involved in analogical reasoning and has been the main focus of research in both Psychology and Artificial Intelligence (Gentner & Forbus, 2011). Theories of analogy have largely converged on the assumption that the process of structural alignment is guided by tacit constraints that lead to the extraction of shared systems of relations with maximal structural consistency. One such constraint is *one-to-one correspondence*: each element of one representation may match with, at most, one element from the other. The second requirement for structural consistency is *parallel connectivity*: if two predicates correspond, their arguments playing like roles must also correspond. Among interpretations that satisfy structural consistency, the cognitive system will prefer those that involve the projection of larger and deeper systems of relations (the *systematicity principle*).

The thesis that people try to satisfy structural consistency during mapping has received considerable empirical evidence. For example, Spellman and Holyoak (1992) asked a group of students at the University of California, Los Angeles, to find out how they interpreted the analogy between the Persian Gulf situation and World War II, launched by President George Bush in a campaign to generate support for a military intervention to free Kuwait from the invasion carried out by Saddam Hussein, the dictator of Iraq. A basic representation of World War II amounted to a story figuring Hitler, the dictator of Germany, and two clear-sighted statesmen, Winston Churchill in Britain and Franklin Delano Roosevelt in the USA. The analogy did not admit one unique interpretation, since competing pressures could conduce to alternative mappings. The match between the USA of 1991 and the USA of World War II would lead to pair Bush to Roosevelt. However, the USA did not enter the war until Hitler had marched through much of Europe. In contrast, Churchill led his nation in early opposition to aggression. Students might interpret then that the USA of 1991 mapped to Britain of World War II and, as a consequence, that Bush should be mapped to Churchill. However, the USA was the major supplier of arms and equipment for the Allies, a role parallel to that played by the USA in the Persian Gulf situation. This opened up a third interpretation, according to which the USA should map to itself, but Bush should map to Churchill.

Nothing prevented students from giving the above mappings as answers, but theories of analogy sustained that people prefer structurally consistent interpretations, that is, those satisfying one-to-one mappings (Bush to either Churchill or Roosevelt, but not to both) and parallel connectivity (keeping leaders and countries together). Results supported this prediction: students tended to provide mappings based on either the two structurally consistent set of correspondences and avoided the inconsistent one.

Clement and Gentner (1991) carried out a series of experiments to test the influence of systematicity on analogical mapping, inference, and evaluation, that is, whether a given predicate is more likely to figure in the interpretation of an analogy if the predicate participates in a common system of relations, and if this interpretation would be preferred. Each of the analogies consisted of a base and target passage describing novel organisms on fictional planets (see Table 1.1 for an example). Each passage included two chief paragraphs. One paragraph described a causal structure that matched between the base and target, and the other described a causal structure that did not match. Subjects had to make mapping choices between *key facts* (effects in the causal structure) that were embedded either in the matching or in the nonmatching causal structures. The key facts themselves always matched between the

**Table 1.1** Example of relational structures used by Clement and Gentner (1991, experiment 1)

| Base: the tams | Target: the robots |
|---|---|
| Consume minerals with underbellies<br>**Exhaust minerals in one spot and must relocate on the rock**<br>*So they stop using underbelly* | Gather data with probes<br>**Exhaust data in one place and must relocate on the planet**<br>*So they stop using probes* |
| Born with inefficient underbelly<br>**Underbelly adapts and becomes specialized for one rock**<br>*So underbelly cannot function on new rock* | Designed with delicate probes<br>***Robots cannot pack probes to survive flight to a new planet***<br>*So probes cannot function on new planet* |

Note. Key facts are shown in italics. Causal information is shown in boldface. Nonmatching antecedent is bolditalic base and the target, but causal antecedents did not. Participants followed the systematicity principle, showing a clear inclination for choosing and evaluating as better those matches that were embedded in a larger system of correspondences. The structure of Experiment 2 was identical to that of Experiment 1, with the difference that the target effects were removed, and participants had to make analogical predictions about the target domain. Subjects predicted information that followed from a causal system that matched the base domain, rather than information that was equally plausible, but that created an isolated match with the base. Results support the idea that mapping, inference, and evaluations concern systems of relations rather than isolated relations.

Even though the analogical subprocesses of mapping, inference, and evaluation may demand the simultaneous consideration of alternative interpretations, studies as the ones reviewed above suggest that people can effortlessly compute them to the extent that the analogs are themselves comprehensible and not overloaded with information (but see Holyoak, Novick, & Melz, 1994, for exceptions to this general rule). The challenge of computational modeling in the field of analogy became then to simulate analogical mapping via computations that are psychologically plausible, that is, attendant to the known competence limitations of working memory.

## 1.3 A Computational Model of Analogical Mapping

To describe the principles, representations, and mechanisms that are implied in the interpretation of an analogy, we will present the structure-mapping theory and its computational model (SME) since it can be considered very representative of the most important theories and computational models in the area. In so doing, we will employ a simple version of the well-known analogy between water flow and heat flow, as it could be presented by a teacher (see Gentner, 1989). We will assume that the learner is presented with the base and target situations in a pictorial format, as in Fig. 1.1.

The professor describes the analogy to the learner as follows: "Heat flow can be understood like water flow. In the water-flow situation (Panel A), the pressure in the beaker is greater than the pressure in the vial, and this causes the flow of water from the beaker to the vial. In the heat situation (Panel B), the temperature of the coffee is higher than the temperature in the ice cube. Try to conjecture why the coffee cools off and the ice cube melts." Besides this complete exposition and precise question, the learner is also aware that the diameter of the beaker is greater than the diameter of the vial, as well as that water and coffee are liquid and have a flat top (see Fig. 1.2 for graph representations of the initial base and target representations of the learner).

The comprehension of an analogy implies finding the correspondences between the elements composing the base and the target, generating interpretations of the analogy, eventually deriving inferences for the target, and, finally, selecting the best interpretation between the available ones. The structure-mapping theory supposes that the analogical machinery follows some implicit rules to limit the space of

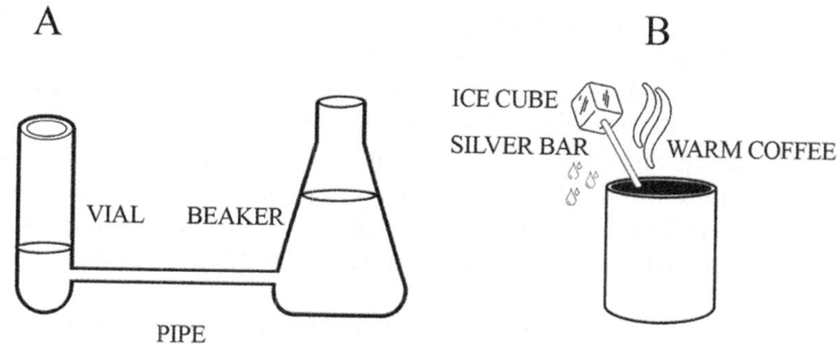


**Fig. 1.1** The analogy between the water flow and the heat flow. Adapted from Gentner (1989, p. 203)

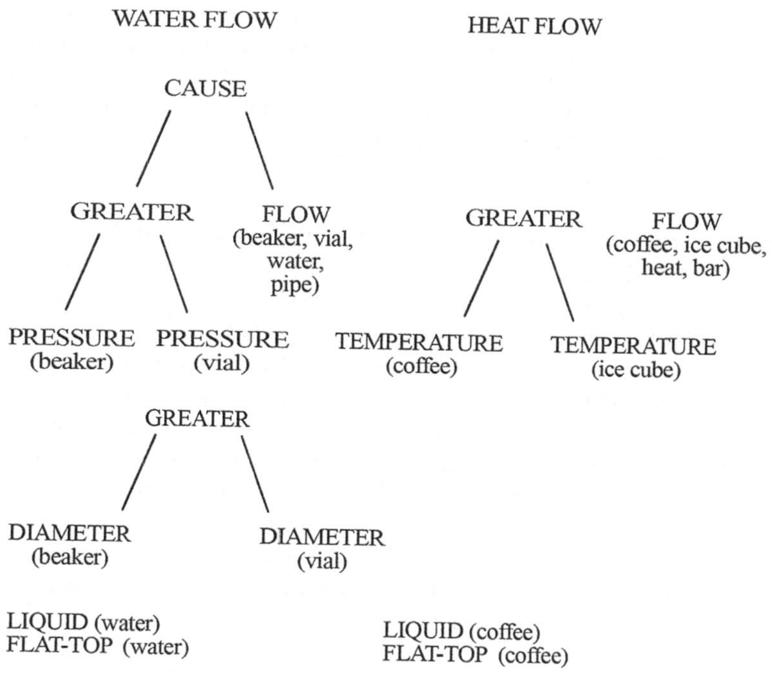


**Fig. 1.2** Graph representations of the initial information taking part of the analogy between the water flow and the heat flow

possible match hypotheses between base and target elements, so as to avoid an intractable combinatorial explosion. We will refer to these rules as *initial conditions*. SME includes three initial conditions: formal identity (relations can only be matched to relations, functions to functions, and entities to entities), identically for relations, and mapping of entities according to matched relations as well as to the roles played in those relations. Once all acceptable element matches have been generated, the system must group them so as to advance in the construction of possible interpretations of the analogy. We refer to the rules that determine which matches can be grouped and which not as *constraints*. The only constraint employed by SME is structural consistency, which can be decomposed, as we saw, in one-to-one mapping and parallel connectivity. Finally, the system must evaluate the available interpretations so as to select the most interesting one (inferences are considered in the evaluation). During evaluation, the analogical machinery prefers to match large, deep-connected systems of relations instead of isolated ones (the *systematicity principle*).

Given base and target descriptions, SME finds globally consistent interpretations via a local-to-global match process, which can be divided into three phases:

***Phase One: Constructing the Match Hypotheses Network*** SME begins by proposing local correspondences by positing all local identity matches between relations in the two representations. Additional matches are made via local parallel connectivity—for example, if two relations are matched, then SME matches their arguments. The initial network is inchoate, providing the material for potential mappings. As SME is applying the rules corresponding to analogy comparisons, attributes (i.e., one-place predicates) will be ignored, and therefore no matches between them will be constructed.

***Phase Two: Parallel Construction of Structurally Consistent Kernels*** At this point, SME starts building mappings by extracting structurally consistent sets from the forest of match hypotheses. To achieve this, SME does two things, again in parallel: (1) It marks as inconsistent those match hypotheses that violate the parallel connectivity constraint and (2) marks as mutually inconsistent pairs of match hypotheses that would violate the one-to-one constraint. It then coalesces the local matches into a set of structurally consistent connected structures, called *kernels*. Each kernel receives a structural evaluation score that arises from assigning a local score to each match hypothesis, as well as from implementing a trickle-down process to propagate evidence downwards from a match hypothesis to match hypotheses between the arguments of the corresponding statements. This provides a local means of implementing the systematicity preference, since match hypotheses that participate in a large matching structure will receive higher scores.

***Phase Three: Constructing Global Mappings*** SME uses a greedy-merge algorithm to combine kernels into one or more global mappings. The basic idea is to start with the largest and deepest kernel (that is, the one with the highest structural evaluation) and serially add others that are structurally consistent with it. This results in one or a few large, structurally consistent global mappings which reveal common structure between the compared representations. At this stage, candidate inferences may be projected from one representation to the other.

If SME were to start from the same base and target representations of our imaginary learner (see Fig. 1.2), it would create match hypotheses between the *greater-than* relations occurring in the base and target. After that, it would generate match hypotheses between their arguments—since both are *functions*[1]—and between the arguments of these functions, since they are entities. At this stage, the program will have a large number of local matches. Figure 1.3 shows the global mappings that are formed from the collection of match hypotheses for this analogy.

After the first merging, only isolated collections stemming from common relations exist. After that, the system combines the pressure to temperature mapping with the flow mapping, since they have a common base structure (the cause predication). Notice that the flow mapping is structurally consistent with the diameter to temperature mapping. However, because the system has placed the flow mapping into the same global mapping as the pressure to temperature mapping, the system is unable to combine the flow mapping with the diameter to temperature global mapping. Each global mapping represents a set of correspondences that can serve as an interpretation of the match. For new knowledge to be generated about the target, there must be information from the base which can be carried over into the target.

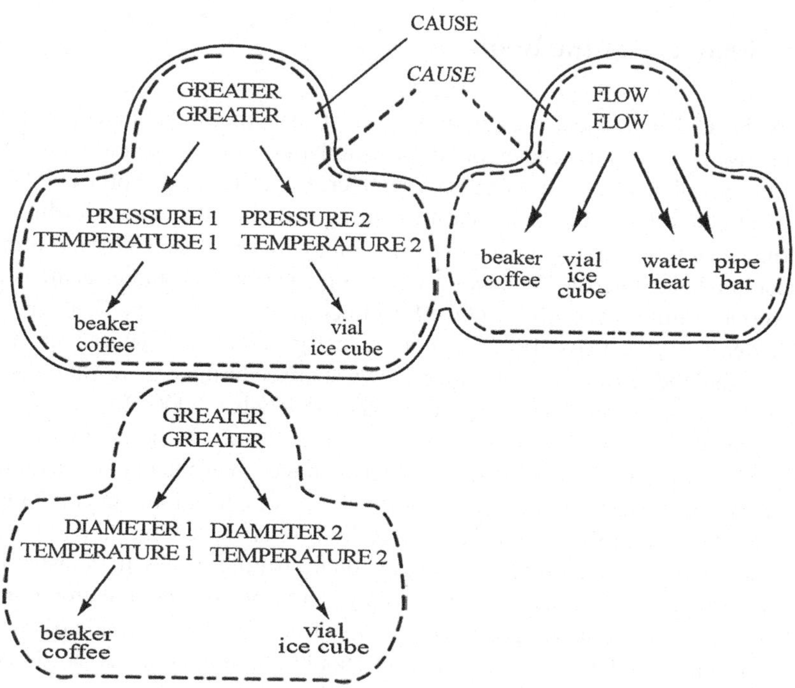


**Fig. 1.3** Global mappings constructed by SME during the interpretation of the heat flow–water flow analogy

[1]Functions map one or more entities into another entity or constant. For example, velocity (car) maps the physical object *car* into the quantity which describes its velocity. Structure-mapping allows substitution of functions to acknowledge their role as an indirect way of referring to entities. All other predicates must be matched identically.

Not just any information can be carried over—it must be consistent with the substitutions imposed by the global mapping, and it must be structurally grounded in it. By structural grounding, we mean that its subexpressions must at some point intersect the base information belonging to the global mapping. In the water flow–heat flow analogy, the inference generated suggests that the difference between the coffee temperature and the ice cube temperature causes the flow of heat from the coffee to the ice cube.

The global matches receive a final structural evaluation, which depends on their local match evidence, the number of candidate inferences they support, and their graph-theoretic structure—for example, the depth of the relational system. In this example, the winning global mapping is the pressure–temperature match discussed above, with its candidate inference of a causal link in the heat domain. The other global mapping contains fewer predicates than the winning global mapping, as well as a shallower relational structure.

## 1.4 Structure of the Book

Researchers of analogical reasoning agree that the process of understanding an analogy is not heavily dependent on surface similarities, that is, similarity of entities fulfilling parallel roles in the analogs (Gentner, Rattermann, & Forbus, 1993). In contrast to the role of surface similarities during mapping, the extent to which they determine the retrieval of base analogs from long-term memory (LTM) has been the subject of a heated debate. Whereas a wealth of experimental studies demonstrates that surface similarity exerts a powerful effect on retrieval (e.g., Gentner et al., 1993; Keane, 1987; Ross, 1989), a more recent generation of naturalistic studies suggests that the retrieval of participants' own sources during real-world tasks does not require surface similarities (e.g., Blanchette & Dunbar, 2000, 2001).

Besides its obvious relevance to the basic question about the nature of human intelligence, the empirical inconsistency between the experimental and naturalistic traditions has important implications for the design of educational interventions. If the naturalistic tradition is right, then only by further inspecting natural environments will educators identify the secret ingredients behind successful transfer. If, on the contrary, it is experimental studies who got it right, then natural environments no longer represent a privileged window into successful transfer, and both experimental results and theory-driven intuitions should retain their status as sources of inspiration for educational practices. In the first part of the present monograph we will flesh out in greater detail the controversy around empirical studies of analogical retrieval. While Chap. 2 will review the main findings and attendant computational simulations arising from the experimental tradition, Chap. 3 will concentrate on the evidence provided by naturalistic approaches. Chapter 4 will compile a set of recent results that we carried out to settle the experimental–naturalistic debate, therefore arriving at a more accurate diagnosis of our ability to retrieve distant analogs from memory. Based on this diagnosis, in the second part of the present monograph we will review what we consider to be the most realistic avenues for honing in on a set of applicable interventions to make distant analogizing more likely.

## References

Black, M. (1962). *Models and metaphors*. Ithaca, NY: Cornell University Press.
Blanchette, I., & Dunbar, K. (2000). How analogies are generated: The roles of structural and superficial similarity. *Memory & Cognition, 28*, 108–124.
Blanchette, I., & Dunbar, K. (2001). Analogy use in naturalistic settings: The influence of audience, emotion, and goals. *Memory & Cognition, 29*, 730–735.
Boden, M. A. (1990). *The creative mind: Myths and mechanisms*. London: Weidenfeld & Nicholson.
Carbonell, J. G. (1981). *Metaphor: An inescapable phenomenon in natural language comprehension*. CMU Technical Report, Department of Computer Science, Carnegie-Mellon University.
Clement, C. A., & Gentner, D. (1991). Systematicity as a selection constraint in analogical mapping. *Cognitive Science, 15*, 89–132.
Dorolle, M. (1949). *Le raisonnement par analogie*. Paris: P.U.F.
Doumas, L. A. A., & Hummel, J. E. (2005). Approaches to modeling human mental representations: What works, what doesn't, and why. In K. J. Holyoak & R. G. Morrison (Eds.), *The Cambridge handbook of thinking and reasoning* (pp. 73–91). Cambridge: Cambridge University Press.
Duncker, K. (1926). A qualitative (experimental and theoretical) study of productive thinking (solving of comprehensible problems). *Journal of Genetic Psychology, 68*, 79–116.
Falkenhainer, B., Forbus, K. D., & Gentner, D. (1989). The structure-mapping engine: Algorithm and examples. *Artificial Intelligence, 41*, 1–63.
Gelb, M. J., & Miller Caldicott, S. (2007). *Innovate like Edison: The five-step system for breakthrough business success*. New York: Penguin.
Gentner, D. (1977a). Children's performance on a spatial analogies task. *Child Development, 48*, 1034–1039.
Gentner, D. (1977b). If a tree had a knee, where would it be? Children's performance on simple spatial metaphors. *Papers and Reports on Child Language and Development, 13*, 157–164.
Gentner, D. (1983). Structure-mapping: A theoretical framework for analogy. *Cognitive Science, 7*, 155–170.
Gentner, D. (1989). The mechanisms of analogical transfer. In S. Vosniadou & A. Ortony (Eds.), *Similarity and analogical reasoning* (pp. 199–242). Cambridge, UK: Cambridge University Press.
Gentner, D., & Forbus, K. (2011). Computational models of analogy. *WIREs Cognitive Science, 2*, 266–276.
Gentner, D., Rattermann, M. J., & Forbus, K. D. (1993). The roles of similarity in transfer: Separating retrievability from inferential soundness. *Cognitive Psychology, 25*, 431–467.
Herbart, J. H. (1898). *Letters and lectures on education*. London: Sonnenschein.
Hesse, M. (1966). *Models and analogies in science*. Notre Dame: University of Notre Dame.
Hofstadter, D. R. (1985). *Metamagical Themas: Questing for the essence of mind and pattern*. London: Viking.
Hofstadter, D. R., & Sander, E. (2013). *Surfaces and essences: Analogy as the fuel and fire of thinking*. New York: Basic Books.
Holland, J. H., Holyoak, K. J., Nisbett, R. E., & Thagard, P. (1986). *Induction: Processes of inference, learning, and discovery*. Cambridge, MA: MIT Press.
Holyoak, K. J. (1984). Analogical thinking and human intelligence. In R. J. Sternberg (Ed.), *Advances in the psychology of human intelligence* (Vol. 2). Hillsdale, NJ: Erlbaum.

Holyoak, K. J., Novick, L. R., & Melz, E. R. (1994). Component processes in analogical transfer: Mapping, pattern completion, and adaptation. In K. J. Holyoak & J. A. Barnden (Eds.), *Advances in connectionist and neural computation theory* (Analogical connections) (Vol. 2, pp. 113–180). Norwood, NJ: Ablex.

Holyoak, K. J., & Thagard, P. (1989). Analogical mapping by constraint satisfaction. *Cognitive Science, 13*, 295–355.

Holyoak, K. J., & Thagard, P. R. (1995). *Mental leaps: Analogy in creative thought*. Cambridge, MA: The MIT Press.

Hummel, J. E., & Holyoak, K. J. (1997). Distributed representations of structure: A theory of analogical access and mapping. *Psychological Review, 104*, 427–466.

Keane, M. T. (1987). On retrieving analogues when solving problems. *Quarterly Journal of Experimental Psychology, 39*, 29–41.

Keane, M. T. (1988). *Analogical problem solving*. Chichester: Ellis Horwood.

Lakoff, G., & Johnson, M. (1980). *Metaphors we live by*. Chicago: Chicago University Press.

Miller, G. A., & Johnson-Laird, P. N. (1976). *Language and perception*. Cambridge: Cambridge University Press.

Ortony, A. (1979a). Beyond literal similarity. *Psychological Review, 86*, 161–180.

Ortony, A. (1979b). *Metaphor and thought*. Cambridge: Cambridge University Press.

Piaget, J., Montangero, J., & Billeter, J. (1977). La formation des correlats. In J. Piaget (Ed.), *Recherches sur l'abstraction reflechissante I* (pp. 115–129). Paris: P.U.F.

Ross, B. H. (1989). Distinguishing types of superficial similarities: Different effects on the access and use of earlier problems. *Journal of Experimental Psychology: Learning, Memory, and Cognition, 15*, 456–468.

Rumelhart, D. E., & Norman, D. A. (1981). Analogical processes in learning. In J. R. Anderson (Ed.), *Cognitive skills and their acquisition*. Hillsdale, NJ: Erlbaum.

Schank, R. C. (1972). Conceptual dependency: A theory of natural language understanding. *Cognitive Psychology, 3*, 552–631.

Schank, R. C., & Abelson, R. P. (1977). *Scripts, plans, goals and understanding: An inquiry into human knowledge structures*. Hillsdale, NJ: Erlbaum.

Spearman, C. (1927). *The nature of "intelligence" and the principles of cognition*. London: Macmillan.

Spellman, B. A., & Holyoak, K. J. (1992). If Saddam is Hitler then who is George Bush?: Analogical mapping between systems of social roles. *Journal of Personality and Social Psychology, 62*, 913–933.

Sternberg, R. J. (1977a). *Intelligence, information processing and analogical reasoning: The componential analysis of human abilities*. Hillsdale, NJ: Lawrence Erlbaum Associates, Inc..

Sternberg, R. J. (1977b). Component processes in analogical reasoning. *Psychological Review, 84*, 353–378.

Ward, T. B., Smith, S. M., & Finke, R. A. (1999). Creative cognition. In R. J. Sternberg (Ed.), *Handbook of creativity* (pp. 189–212). Cambridge: Cambridge University Press.

Winston, P. H. (1980). Learning and reasoning by analogy. *Communications of the ACM, 23*, 689–703.

# Part I
# Assessing Our Ability to Retrieve Analogous Situations

# Chapter 2
# The Experimental Tradition

Educators have long been concerned about students' alleged difficulties for transferring knowledge to contents and contexts different from those of the initial learning. Known as "the problem of inert knowledge," these difficulties still remain at the center of educational debate (Barnett & Ceci, 2002; Day & Goldstone, 2012). The ability to establish distant analogies represents a powerful tool for exporting contents across thematic barriers. The theoretical and computational model reviewed in Chap. 1 reflects the finding that the central processes of mapping and inference are not particularly taxing: When the representations of the analogs are neither too large nor saturated with irrelevant information, both humans and computer simulations rapidly hone in on the right set of correspondences. But if the mapping and inference stages are rather straightforward, where do the transfer difficulties experienced by students originate? One conspicuous suspect is *analogical retrieval*, that is, the process of being recalled of a potentially useful base analog on the basis of a target analog that is currently being processed.

Human memory takes advantage of distinct short-term and long-term storages. As with the RAM and ROM modules of our computers, our working memory contrasts sharply with our long-term memory (LTM) in aspects as crucial as the duration of the information and the easiness with which such information is retrieved. Just as ROM modules (hard-drive disks), our LTM has an impressively large capacity that enables us to learn an endless number of concepts and facts, as well as to be able to recall them after long delays. Its most relevant downside has to do with the rather clumsy way in which this information is accessed. While some contents are quite straightforwardly retrieved, other contents require time and effort to be eventually located, with the obvious risk of not being able to recall them at all. The net capacity of RAM modules is ridiculously smaller than that of ROM, but it can afford the "reading and writing" of information at mesmerizing speeds. In a similar fashion, our short-term memory supports super-fast processing of the contents of our flow of consciousness, whose maximum size has been traditionally estimated in



M. Trench, R. A. Minervino, *Distant Connections: The Memory Basis of Creative Analogy*, SpringerBriefs in Psychology,
https://doi.org/10.1007/978-3-030-52545-3_2

as few as 7 ± 2 unrelated items. It would seem that solving a quotient with decimal digits demands simultaneously handling much more than seven items. But it is the blackboard or a sheet of paper what is actually working as a transient extension of our working memory. Just attempt this same mental operation without any of these physical prostheses and experience how laborious the reading and writing on LTM can become.

From the conformation of our memory systems it follows that working memory could conceivably cope with maintaining a base and a target analog simultaneously active, as when a professor cues us to understand the structure of the atom in terms of that of the solar system. But what would be the perspectives for creatively connecting a novel situation with a structurally similar situation that is buried in LTM along with zillions of spurious or irrelevant situations? Some theorists have contended that, in contrast to the feasibility of scanning LTM for representations involving specific concepts—think of the speed with which Google searches the entire web on the basis of particular keywords—the possibility of performing full-fledged structural comparisons between the target and all representations stored in memory is computationally unrealistic. Hence, human reasoners should seldom retrieve analogous situations whose individual constituents do not overlap, at least in part, with those of the target.

In the last decades, a considerable body of empirical research was carried out to assess the severity of this limitation. As in the study of memory at large, the experimental procedure most widely used to investigate analogical retrieval comprises two phases. During the learning (or encoding) phase, participants receive one or several base analogs interleaved among distracter situations. During the retrieval phase, participants receive the target analog and the experimenters assess whether its processing leads to the retrieval of the critical situation learned during the previous phase. This two-phase paradigm allows the experimenters to manipulate the conditions under which the base and the target analogs are processed, thus shedding light on diverse factors that facilitate (or hinder) human access to analogically related material. The majority of experimental studies on analogical retrieval can be classified in terms of the encompassing task in which the target analog is embedded. The following section will review a series of behavioral experiments aimed at investigating analogical retrieval during problem-solving activities. Next we will review a parallel series of experimental studies employing cued-recall tests of story reminding. The remainder of the chapter will discuss the computational models that were created to simulate the converging results obtained with both types of studies.

## 2.1 Experimental Studies of Analogical Problem-Solving

Problem-solving is a broad category that comprises a wide range of activities in which one aims at achieving a desired state from an initial state of affairs and a set of permitted actions (operators) and limiting factors (restrictions, Newell & Simon, 1972). Taxonomies of problems abound, and focus on factors as diverse as whether or not the above components are clearly defined (well-defined vs. ill-defined problems), the amount of prior knowledge needed (knowledge-rich vs. knowledge-lean problems), or the temporal nature of the solution process (sequential, stepwise solutions vs. sudden, insightful solutions). While Reed, Ernst, and Banerji's (1974) study with well-defined, sequential problems should be considered the first major study that documented the pitfalls of analogical problem-solving, it was Gick and Holyoak's (1980) study with ill-defined problems which gave rise to a rapid and cohesive succession of studies scrutinizing the variables that govern successful analogical access.

Along a series of experiments, Gick and Holyoak (1980) assessed their participants' ability to transfer solution procedures across thematically distant situations. In the fourth experiment of the series, students were told that they were going to participate in a study comprising two unrelated sections: one in which they would be asked to memorize stories, and another one in which they would be given problems to solve. During the first, participants received a story describing a problem and its solution, together with two distractor stories. After studying each story for 15 min, participants were asked to reproduce them with as much detail as possible. After a 5-min break, participants were asked to propose one or more solutions to Duncker's (1945) *Radiation problem*:

> Suppose you are a doctor faced with a patient who has a malignant tumor in his stomach. It is impossible to operate on the patient, but unless the tumor is destroyed the patient will die. There is a kind of ray that can be used to destroy the tumor. If the rays reach the tumor all at once at a sufficiently high intensity, the tumor will be destroyed. Unfortunately, at this intensity the healthy tissue that the rays pass through on the way to the tumor will also be destroyed. At lower intensities the rays are harmless to healthy tissue, but they will not affect the tumor either. What type of procedure might be used to destroy the tumor with the rays and at the same time avoid destroying the healthy tissue?

One of the three stories read during the previous phase had been especially crafted by the experimenters to suggest a potential solution to the radiation problem:

> A small country fell under the iron rule of a dictator. The dictator ruled the country from a strong fortress. The fortress was situated in the middle of the country, surrounded by farms and villages. Many roads radiated outward from the fortress like spokes on a wheel. A great General arose who raised a large army at the border and vowed to capture the fortress and free the country of the dictator. The General knew that if his entire army could attack the fortress at once it could be captured. His troops were poised at the head of one of the roads leading to the fortress, ready to attack. However, a spy brought the general a disturbing report. The ruthless dictator had planted mines on each of the roads. The mines were set so that small bodies of men could pass over them safely, since the dictator needed to be able to move troops and workers to and from the fortress. However, any large force would detonate the mines. Not only would this blow up the road and render it impassable, but the dictator would destroy many villages in retaliation. A full-scale direct attack on the fortress therefore appeared impossible. The General, however, was undaunted. He divided his army up into small groups and dispatched each group to the head of a different road. When all was ready he gave the signal, and each group charged down a different road. All of the small groups passed safely over the mines, and the army then attacked the fortress in full strength. In this way, the General was able to capture the fortress and overthrow the dictator.

The military story and the radiation problem share a similar system of relations and roles, albeit at a high level of abstraction. In both cases, certain force can potentially eliminate a central target if applied at a given intensity, but with the consequence of harming other important elements that need to be preserved. As in most analogies, this isomorphism breaks down at lower levels of abstraction: for example, while high-intensity rays would simply destroy tissues, the passage of the full army would detonate the mines, which would in turn destroy the roads plus the army itself. Despite violating the principle of one-to-one mapping, acknowledging a parallelism between both situations allows exporting the base solution to the target problem: just as the General had smaller groups converge on the fortress through different roads, so could weaker rays be pointed to the tumor from different locations, such that they only add up at their intersection.

Gick and Holyoak (1980) compared the frequency of convergent solutions (i.e., those wherein weaker beams are pointed to the tumor from different locations) under two experimental conditions: with vs. without an explicit indication to consider the usefulness of the stories read during the previous session. The contrast between these conditions was very sharp: Whereas prompted participants transferred the convergent solution in 75% of the cases, only 30% did so in the lack of this indication (the net proportion of participants who spontaneously retrieved the base situation should be considered to be even lower, if one takes into account the fact that an independent group of participants who had not received the military story still proposed convergent solutions in 10% of the cases). This gap could be easily interpreted in terms of the different psychological restrictions entailed by comparing the target analog against the whole of LTM vs. a small set of candidates within which the critical analog is contained: while prompted participants were potentially able to select the three base stories *one-by-one* for a structural check against the target in working memory, unprompted participants would have had to perform a serial search within a memory set that is orders of magnitude larger.

But the picture could be even more dramatic than Gick and Holyoak's (1980) results suggest. Spencer and Weisberg (1986) suspected that the small proportion of participants who proposed convergent solutions without any prompt could have taken advantage of a rather weak contextual separation between the learning and the transfer phases. They argued that in real life it would be unlikely that the context in which the target analog is encountered coincides with the context in which the source analog has been encoded, especially in those cases where the analogs belong to thematically unrelated domains. In order to increase the contextual separation between the learning and the transfer phases, Spencer and Weisberg asked permission from a Psychology instructor to run a pilot study on text comprehension, wherein participants were required to answer comprehension questions about the military story plus another two distractor stories. Once the task was completed and the experimenters had left the room, the course instructor presented participants with the radiation problem in the context of a class demonstration on insight problem-solving. The proportion of convergent solutions (12%) was even lower than that of Gick and Holyoak and, more critically, it was not statistically different from the proportion of convergent solutions proposed by another group of participants who had not been previously exposed to the military problem (4%). These results gave further support to the thesis that when structural similarity is not accompanied by object similarity, the spontaneous retrieval of source analogs is very unlikely.

By means of a procedure similar to that of Spencer and Weisberg (1986), Keane (1987) set forth to investigate the extent to which the presence of surface similarity can augment the probabilities of spontaneously retrieving a source analog from memory. In the interdomain condition, the instructor of a course on Cognitive Psychology presented participants with the military story as part of a classic demonstration of memory research. Students had to study it for 4–5 min, and to recall it half an hour later with as much detail as they could. In the intradomain condition, participants received a story that was both structurally and superficially similar to the radiation problem: it told about a surgeon who was attempting to use a type of ray to destroy a cancer located at the center of a patient's brain. While his rays would destroy the tumor if applied at a high intensity, at such an intensity they would also destroy brain tissue. Akin to the solution applied in the military story, the surgeon divided his rays up into batches of low-intensity rays, and had them converge on the cancer from a number of different directions. Near the end of the lesson, the experimenters entered the class and recruited participants for a study on problem-solving, which was individually administered during the three subsequent days. Prior to receiving the radiation problem, participants read an instructional material about the usefulness of analogies for solving problems, which included examples of the role of distant analogies in scientific discovery. Then participants received the radiation problem plus two distracter problems, and were asked to think of analogous problems before trying to solve each of the problems, verbalizing any idea that came to mind. Results showed a strong effect of surface similarities: while the interdomain base analog was retrieved only 12% of the times, the intradomain story was retrieved in 88% of the cases.

In a subsequent experiment, Keane (1987) assessed whether a medical cover story was required for retrieval, as well as whether matching elements needed to be identical or simply similar. To this end, the intradomain condition of Experiment 1 was replaced by two interdomain conditions: one in which a central element (rays) was identical to that of the radiation problem, and another in which it was only similar (lasers). The stories stated that a General wanted to destroy an intercontinental missile using certain type of rays (laser beams) which would be entering the atmosphere at great speeds. The general needed to use a high-intensity ray (high-intensity laser beam) in order to destroy the incoming missile, but such high-intensity rays (high-intensity laser beams) were inaccurate because they would heat up the air through which they passed, thus distorting their trajectory. The stories ended by stating that the General divided the high-intensity ray (high-intensity laser beam) into a number of low-intensity rays (low-intensity laser beams). By sending them simultaneously from different locations, they converged on the missile making up a sufficiently powerful beam to destroy it. Results revealed that the retrieval of these two stories (55%) was much higher than that of the original military story (5%), suggesting that it is not necessary that the source analogs belong to the same thematic domain as the target: a few superficial similarities suffice for reliable analogical retrieval to occur. Furthermore, the lack of differences between the retrieval rate of the rays' version (58%) and that of the laser beams' version (55%) demonstrated that it is semantic similarity and not only semantic identity what drives analogical retrieval. One question raised by the above results concerns whether the semantic similarity between base and target elements should be between base and target elements playing corresponding roles across analogs. Ripoll (1998) demonstrated that the effect of semantic similarity on retrieval operates uniformly regardless of whether the objects maintaining semantic similarity occupy parallel roles in their corresponding structures.

While successful transfer can be taken as a proof that the base analog has been retrieved, the question remains as to whether the lack of successful transfer can be taken as evidence that the critical source situation had failed to be retrieved from LTM. It is in principle conceivable that participants may activate the representation of the base situation in working memory, but still fail to acknowledge a connection with the target. To assess this possibility, Anoli et al. (2001, Experiment 3) had participants read a convergence story in which the initial idea of filling an artificial lake by channeling a great stream of water through a wide canal was anticipated to produce seasonal overflows, which could damage surrounding areas. For this reason, the original plan was later replaced by the use of concurrent smaller canals, which could collectively carry the same flow of water but without the risk of overflowing. During a subsequent phase, two groups of participants were asked to solve the *Lightbulb problem*, an analog of the radiation problem in which a laser beam of the intensity that was required to fuse the broken filament of an expensive lamp would also melt the crystal bulb of the lamp. In one of the conditions, the experimenters interrupted participants' efforts to find a solution to the target problem by asking them to state what had been the main drawback of the initial strategy for filling the pond in the previous story, as well as how it was overcome by the strategy that was finally employed.[1] Quite strikingly, the incidental activation of the source analog during the processing of the target did not increase the proportion of convergent solutions, as compared to the standard condition in which the base story was not brought to participants' attention while solving the target.

To summarize, a tightly connected set of studies using variants of the military situation and the tumor problem converge in demonstrating that the spontaneous retrieval of analogous situations is highly infrequent in the absence of superficial similarities between the base and the target. They show, however, that retrieval does not require a great deal of superficial overlap to occur. When a shared feature is distinctive, as is the case with the added presence of rays/beams in Keane's (1987) adaptation of the military story, it may suffice to facilitate the retrieval of situations bearing a similar structure.

[1] Needless to say, participants could have treated the experimenters' intrusion quite differently from the sudden reminding that spontaneously pops up in their mind. While the former could legitimately be interpreted as an unrelated interruption, the latter may be interpreted as bearing some meaningful connection to the target episode being processed.

## 2.2 Experimental Studies of Story Reminding

Studies of analogical problem-solving revealed the difficulties experienced by people to recall a distant base analog, even when there isn't any literally similar experience to draw upon. The fact that the only source available in LTM fails to be retrieved in such a catastrophic way represents a prototypical example of the problem of inert knowledge, since a student has exactly what is needed to solve a problem, but the presence of surface mismatches prevents her from noticing its relevance. In a manner similar to the solution of typical insight problems, a seemingly absurd mental limitation blocks access to the solution.

Though counterintuitive from a purely educational standpoint, this drop-off of analogical retrieval in the absence of semantically similar elements between the problems could almost have been predicted based on the complexity involved in performing a full structural mapping between the target and each of an overwhelming number of candidate situations stored in LTM. On top of the intrinsic computational challenge of finding a needle in a haystack, the fact that satisfactory solutions to a problem usually clarify important aspects of the very nature of the problem would entail that participants' initial representation of the target problem could be somewhat suboptimal. To exemplify this later point, it is far from clear that participants would spontaneously represent the radiation available to the surgeon in a straight-line, laser-like fashion that would match the directionality of the army, let alone the fact that rays would *add up to*, rather than *cancel*, each other.

One question of interest is whether the retrieval of distant analogs would still be difficult in situations where, as opposed to the typical case in analogical problem-solving, participants (a) know that the critical source analog lies within a well-defined subset of LTM and (b) are potentially able to build more complete representations of the target analogs than in the case of unsolved problems. The studies of story reminding to be reviewed in the present section had the potential to illuminate these issues.

In these studies, experimenters first created sets of story pairs differing systematically in their similarity class. In a first phase of the procedure, participants read a large set of stories within which the critical base item is buried among several distracter items. In a subsequent phase, they read a second set of stories with the task of reporting, for each of the stories of the new set, which story or stories from the original set the new story reminds them of. Taking target stories as a starting point for generating their materials, Gentner, Rattermann, and Forbus (1993) designed the base stories of each set as combinations of different levels of predicate matches. Matches could occur at three levels of predicates: (1) one-place predicates representing properties of entities; (2) first-order relations, that is, relations between entities (e.g., X shooting at Y/P firing at Q); and (3) higher-order relational structure, such as causal relations that determine whether two events pertain or not to particular plot structures. These three levels of matches—object properties, first-order relations, and higher-order relational structure—were combined to yield different kinds of base stories: *literal similes*, which resemble the target in object properties, first-order relations, and higher-order relations, *analogies*, which match the target in first-order and higher-order relations, *mere-appearance matches,* which match the target in object properties and first-order relations, and *FOR matches*, which match only in first-order relations. Table 2.1 presents the most widely known set of stories from Gentner et al.'s studies.

Averaging across experiments from Gentner et al. (1993), the retrieval of analogical matches (21%) was rather low, comparable to the retrieval rates of the problem-solving tradition. Hence, the difficulty of retrieving distant analogs obtained in problem-solving studies seems to be replicated even when people have episodic clues that could potentially narrow the search space down to a set of items that, albeit large enough to impede a serial type of search, represents a minimal fraction of LTM. Moreover, it was replicated using close-ended stories whose structure could be readily captured. Besides replicating and generalizing the challenge of retrieving distant analogs from memory, Gentner et al.'s results revealed that the probabilities of retrieving distant analogs are lower than those of other kinds of similarity matches. While it was not surprising that distant analogs got less retrieved

**Table 2.1** Example of the sets of source and target stories used by Gentner et al. (1993)

| |
|---|
| Story |
| Karla, an old hawk, lived at the top of a tall oak tree. One afternoon, she saw a hunter on the ground with a bow and some crude arrows that had no feathers. The hunter took aim and shot at the hawk but missed. Karla knew the hunter wanted her feathers so she glided down to the hunter and offered to give him a few. The hunter was so grateful that he pledged never to shoot at a hawk again. He went off and shot deer instead |
| Literal similarity match |
| Once there was an eagle named Zerdia who nested on a rocky cliff. One day she saw a sportsman coming with a crossbow and some bolts that had no feathers. The sportsman attacked but the bolts missed. Zerdia realized that the sportsman wanted her tail feathers so she flew down and donated a few of her tail feathers to the sportsman. The sportsman was pleased. He promised never to attack eagles again |
| Analogy match |
| Once there was a small country called Zerdia that learned to make the world's smartest computer. One day Zerdia was attacked by its warlike neighbor, Gagrach. But the missiles were badly aimed and the attack failed. The Zerdian government realized that Gagrach wanted Zerdian computers so it offered to sell some of its computers to the country. The government of Gagrach was very pleased. It promised never to attack Zerdia again |
| Surface similarity match |
| Once there was an eagle named Zerdia who donated a few of her tail feathers to a sportsman so he would promise never to attack eagles. One day Zerdia was nesting high on a rocky cliff when she saw the sportsman coming with a crossbow. Zerdia flew down to meet the man, but he attacked and felled her with a single bolt. As she fluttered to the ground Zerdia realized that the bolt had her own tail feathers on it |
| First-order relations (FOR) match |
| Once there was a small country called Zerdia that learned to make the world's smartest computer. Zerdia sold one of its supercomputers to its neighbor, Gagrach, so Gagrach would promise never to attack Zerdia. But one day Zerdia was overwhelmed by a surprise attack from Gagrach. As it capitulated, the crippled government of Zerdia realized that the attacker's missiles had been guided by Zerdian supercomputers |

than literal similes, it was somewhat alarming that they were less retrieved than mere-appearance matches (61%).

Wharton, Holyoak, and Lange (1996) provided further evidence about the relative retrievability of different types of story matches. Table 2.2 shows an example of the materials used in this study. The stories in the left column reinstantiate the structure of the "sour grapes" fable wherein the main character fails to attain some goal and then tries to make that failure seem less important by disparaging that goal.

The stories in the right column instantiate a different theme, "self-blame," wherein the main character also fails to attain some goal, but blames his failure on a personal shortcoming. Hence, combinations of a target and a source story drawn from different columns yield pairs of stories that describe different themes: while the source story about "John" (Theme 1) and the literally similar target story about "Jennifer" (Theme 2) both describe characters who fail to achieve a goal, the failure attributions are

**Table 2.2** Example of the sets of materials used by Wharton et al. (1996)

| Theme 1: sour grapes | Theme 2: self-blame |
|---|---|
| Source: John was very confident about himself. He did a lot of homework in order to get good marks. John had only a B+ GPA in his first year in high school. He was sure he could do better. Earlier, a counselor had arranged for him to meet with the recruiter from Yale. When he got home from class, he opened the thin rejection letter from Yale. That night he mentioned to his father how he believed that people from Ivy League schools were pretentious | Source: Derrick had failed to make the gymnastics team last fall. He practiced a lot in order to make the team. He wanted to try again. Derrick was positive he had a lot of potential. His PE teacher had gotten him a tryout with the gymnastics team coach. The gymnastics team coach watched him perform and then told his PE teacher that he didn't want him on the team. Derrick confessed to his teacher that the coach undoubtedly thought he, Derrick, didn't have the talent for gymnastics |
| Close analogous target: Lisa spent long hours trying to make her corporation successful. She was very sure about herself. Lisa had broken up with her fiance a year ago. She wanted to meet someone new. A co-worker set her up to go out with someone he knew well. She waited at the fancy restaurant until 8:30 and then left without ordering dinner. She told her friend that she thought that her date probably wasn't that handsome and that investment bankers are really boring, anyway | Close analogous target: Jennifer worked hard attempting to create a new business venture. She had divorced her husband some time before. She wanted to start socializing again. A friend fixed a blind date for her with one of his friends, Henry, from work. Jennifer was very excited. She waited alone at the entrance of the museum for 2 h. She confessed to her friend that her date thought she wasn't that attractive and that software engineers aren't interesting |
| Distant analogous target: Elle was a unicorn who wanted to see what was on the other side of the river. She thought the lands over there were enchanted and rich with meadows and fruit trees. One day she set out to cross the river. Unfortunately, the water was very fast and too deep. Elle swam as hard as she could but after 20 min she had to turn back because of fatigue. Elle decided that the stories about the land on the other side of the river were just false rumors and that there was probably nothing of worth over there | Distant analogous target: Jane was a unicorn who wanted to see what was on the other side of the river. She thought the lands over there were enchanted and rich with meadows and fruit trees. One day she set out to cross the river. Unfortunately, the water was very fast and too deep. Jane swam as hard as she could but after 20 min she had to turn back, exhausted. Jane decided that she wasn't worthy of being in the magic lands |

somewhat different. As in Gentner et al. (1993), the retrieval rates of distant analogs (20%) was lower than those of literal similes (50%) and mere-appearance matches (40%). In a second experiment, they assessed the retrieval of close/distant analogs/disanalogs across three different study-test delays: 5 min, 24 h, and 7 days. Results revealed that the retrieval patterns of all conditions remained quite stable after long delays. Collapsing across levels of delay, retrieval rates were 70% (literal similes), 30% (mere-appearance matches), and 35% (distant analogs).

Inspired by the procedure, materials, and results of Wharton et al.'s (1996) studies on story reminding, Raynal, Clement, and Sander (2017) called into question the traditional wisdom that retrieval is mostly driven by surface similarity. The main reason for challenging such a widespread assumption stems from their objection—formerly posed by Keane (1987, p. 105)—that the story pairs allegedly having only surface matches in prior studies of story reminding also involved a significant amount of structural overlap. Take as an example the Karla the Hawk set shown in Table 2.1. Despite a number of structural mismatches, in both the target and its corresponding mere-appearance match there is a bird that donates some of its feather to an archer, in the hopes that by so doing the archer would stop aiming at it. The FOR match, in which a small country agreed to sell its supercomputers to its warlike neighbor in the hopes that it would never attack, also conforms to a schema wherein an entity gives something valuable to a dangerous agent in the hopes that such action would dissuade the agent from attacking. While the different endings of the stories yield completely different gists, their shared components represent a conceptual core whose complexity and specificity are on par with those of analogically related items used in other studies of analogical retrieval. In order to avoid this confounding, Raynal et al. (2017) placed a superficially dissimilar analogous source in competence against a superficially similar item bearing no structural overlap with the target story. The target told of an ambulant pizzaiolo who held a pizza truck in a popular place, and reacted to the inconvenient settlement of another pizzaiolo in a nearby location by giving him generous advice about how to improve his dough. The target story ends that in order to show the former pizzaiolo how much he found his intention was nice, the second pizzaiolo relocated his truck to avoid competing with him. While the superficially dissimilar source analog embedded this same abstract structure in the context of two girls who compete for attention from a boy to whom they are both attracted, the superficially similar disanalog told of a food truck called "At Alessandro and Fabio's," whose clientele were fond of the authentic atmosphere steaming from this stand, held by two happy looking pizzaioli dressed in traditional Italian suits. However, the story ends that once the two pizzaioli had left this selling space, they switched to traditional German clothes for selling sausage specialties at "Hans and Hendrich's."

In two experiments, Raynal et al. (2017) found that the retrieval of superficially dissimilar analogs (80%) was nearly four times higher than that of superficially similar disanalogs, a pattern that stands in sharp contrast with those obtained by Gentner et al. (1993) as well as Wharton et al. (1996). How to explain this reversal? One of the reasons provided by the authors points to the familiarity of their materials: "Whereas most studies have focused on abstract interdomain analogies that a novice could rarely access, we aim at investigating analogies between situations inspired from social scenarios that can be experienced in different domains of daily-life" (p. 2959). But take into account the parallelism between the plots of the pizzaioli and the Karla the Hawk story:

A bird is attacked by a hunter « A pizzaiolo is threatened by a potential competitor
Bird spots a defect in hunter's arrows « Pizzaiolo spots a defect in competitor's dough
Bird helps the hunter fix his arrows « Pizzaiolo helps competitor improve his dough
Hunter leaves and attacks other animals « Competitor moves away to avoid competition

If the gists instantiated with anthropomorphic animals are less comprehensible that their realistic human equivalents, then story makers from Aesop to Disney or Pixar have built an empire based on a blatantly wrong assumption, and Raynal et al.'s finding would bear much broader implications than those the (narrow) field of analogical retrieval can humbly aspire.

The original explanation provided by the authors seems somewhat more convincing. In contrast with the mere-appearance matches of the Gentner et al.'s (1993) studies, which share non-negligible amounts of structure with their corresponding cues (see example above), the similarity between the pizzaioli story and that of Hans and Heinrich involves objects (two pizzaioli and an Italian food truck) but not relations. Hence, the high levels of allegedly "surface based" retrieval in Gentner et al.'s and Wharton et al.'s (1996) studies (unlike Gentner et al., Wharton et al. openly recognize that their remote analogs have moderate degrees of structural overlap) could have conceivably been the result of surface as well as structural similarities. While we sympathize with Raynal et al.'s (2017) claim that structural overlap should be eliminated from superficially similar disanalogs, we contend that their attempt to remove structural similarity inconveniently entailed removing any overlap at the level of first-order relations, whose presence in Gentner et al.'s (1993) mere-appearance matches was *definitional*. Rather than being an improved version of mere-appearance matches, Raynal et al.'s superficially similar disanalog should be considered an equivalent to Gentner et al.'s (Experiment 3) *objects-only matches*, whose retrieval rate was low and no different from that of superficially dissimilar analogs in Gentner et al.'s original study. Given that mere-appearance matches are more similar to their corresponding target than objects-only matches, a proper way of determining whether purely structural isomorphs are outcompeted by superficially similar disanalogs should distinguish between these two types of similes.

Even if one were to assess how distant analogs fare against objects-only matches, more careful attention should be drawn to the underlying metric along which the surface and the structural similarities are manipulated. As a contrasting case, consider the independent demonstration of the effects of either surface *or* structural similarity, as in Keane (1987) or Wharton et al. (1996). The manipulations involved in these demonstrations only require two different levels (e.g., high vs. low, high vs. moderate, or some vs. none) of either surface or structural similarity, thus allowing for various degrees of freedom as to how to implement the manipulation. But when pitting one similarity against each other, there needs to be a more principled way of ensuring that the degree of structure in the structural source roughly equals the degree of surface in the superficial cue. To be more concrete, imagine that we wanted to generate a close disanalog for the pizzaioli target whose superficiality would match the degree of structural similarity shared between said target and the flirtatious girls' scenario. Should we include just one similar object, two/three elements—as in Hans and Heinrich's story—or a near-ceiling proportion of similar objects? It is in principle conceivable that the presence of a higher proportion of similar objects might render a story more retrievable than a distant analog, even in the lack of similar first-order relations.

As stated at the beginning of this chapter, our educationally oriented preoccupation about the factors that contribute to the problem of inert knowledge made us pay special attention to the retrieval of distant analogs. But to the extent that distant analogs lack surface similarity altogether, why bother with extensive excursions into the role of surface similarity in retrieval? One of the reasons for being concerned with the strength of surface similarity has to do with the competitive nature of human memory. If surface similarity in fact dominates memory retrieval, then the presence of spurious matches in LTM could complicate access to relevant sources that could perhaps be retrieved in the absence of such matches. The second reason for being concerned with the strength of surface similarity relates to the fact that if our system retrieved non-analogical matches too often, our cognitive system would be frequently disturbed by useless reminding. However, not all surface matches are equally useless: If our memory systematically brought objects-only matches, the situation would be more troublesome than if it brought mere-appearance matches like de ones employed by Gentner et al. (1993). Taking as an example of target analog the story of the Italian pizzaioli, what kind of rubbish would naturally pop up in our consciousness? Will we be reminded of just any episode featuring an Italian pizzeria (e.g., Spike Lee's movie *Do the Right Thing,* wherein three Italians have a pizzeria in Brooklyn), or only superficially similar situations carrying along a significant amount of structure despite embodying a dissimilar gist (e.g., a mere-appearance match wherein a food truck was invaded by another food truck, but they negotiated that they would be open at different times of the day)? In spite of some mismatching features between mere-appearance matches and the actual targets, they could sometimes be useful under particular pragmatics (e.g., arguing via contrasting cases), something that objects-only matches would hardly afford. We will return to this issue in further chapters, when discussing whether the surface bias can be considered a friend or a foe. Before doing so, in the remaining of the chapter we will review the computational models built to simulate the results from the experimental tradition. The second reason for being concerned with the strength of surface similarity relates to the fact that if our system retrieved non analogical matches too often, our cognitive system would be frequently disturbed by useless remindings. However, not all surface matches are equally useless. If our memory systematically brought objects-only matches, the situation would be more troublesome than if it brought mere-appearance matches like de ones employed by Gentner et al. (1993). Taking as an example of target analog the story of the Italian pizzaioli, what kind of rubbish would naturally pop up in our consciousness? Will we be reminded of just any episode featuring an Italian pizzeria (e.g., Spike Lee's movie Do the Right Thing, wherein three Italians have a pizzeria in Brooklyn), or only superficially similar situations carrying along a significant amount of structure despite embodying a dissimilar gist (e.g., a mere-appearance match wherein a food truck was invaded by another food truck, but they negotiated that they would be open at different times of the day)? In spite of some mismatching features between mere-appearance matches and the actual targets, they could sometimes be useful under particular pragmatics (e.g., arguing via contrasting cases), something that objects-only matches would hardly afford. We will return to this issue in further chapters, when discussing whether the surface bias can be considered a friend or a foe. Before doing so, in the remaining of the chapter we will review the computational models built to simulate the results from the experimental tradition.

## 2.3 A Computer Model of Similarity-Based Retrieval: MAC/ FAC

The retrieval patterns obtained in experimental studies inspired the construction of several algorithms engineered to simulate similarity-based reminding, arguably the most significant bottleneck in analogical reasoning.

Gentner et al. (1993) reasoned that any appropriate model of similarity-based retrieval should be strongly but not solely influenced by surface similarity, and somewhat but not wholly insensitive to structural consistency. It should typically retrieve literally similar matches, often retrieve mere-appearance matches, and occasionally retrieve purely analogous matches. As with other contemporary models (e.g., ARCS, Thagard, Holyoak, Nelson, & Gochfeld, 1990), MAC/FAC (many are called but few are chosen, Forbus, Gentner, & Law, 1995) uses a two-stage retrieval process. The first stage is a "wide-net" stage in which a crude, computationally cheap match process is used to pare down the vast set of memory items into a small set of candidates for more expensive processing. The second stage is a structural matcher, namely SME in *literal similarity* mode.

As reviewed in the previous chapter, SME operates by building intermediate structures, in the form of clusters of local matches. The idea of building such networks for a pair of items, or a small number of pairs of items, is psychologically plausible, because the size of the match hypothesis network is polynomial in the size of the descriptions being matched. This means that a fixed-size piece of hardware could be built which could be dynamically reconfigured to represent any cluster of local matches for input descriptions of some bounded size. However, it is implausible that such clusters could be built between the target situation and every item in a large memory pool, and especially that this could happen quickly enough to account for observed retrieval times. Thus, the initial process (i.e., the MAC stage) is meant to be a cheap, fast process whose objective consists in selecting a manageable number of possible analogs to pass along to the FAC stage. In order to gauge a quick estimate of the overlap between the target analog and each item in the memory pool, the MAC stage extracts a content vectors out of the target representation and compares it against the content vectors of all stored representations. Content vectors, which are flat summaries of the knowledge encoded in complex relational structures, are generated by assigning a position in an ordered series to all concepts in LTM, and counting how many times each predicate concept appears in each of the stored situations. For example, if MLP comprises 50 situations collectively featuring 250 predicates, the vectors of each source and target analog would have 250 terms, most of which will have value zero. Upon calculating the dot products between the content vector of the target and the vectors of all situations in LTM, the MAC stage submits to the FAC stage the source item yielding the highest dot product, as well as every other source situation within 10% of it. As larger structures could inflate dot products in a distortive manner, content vectors are normalized such that a fixed amount of weight is distributed among the non-zero fields of the vector, in proportion to the numerosity of each predicate in the sequence.

The FAC stage is responsible for the structural alignment, interpretation, and evaluation of the matches submitted by the MAC stage. Hence, it operates on the structural representations and not on the content vectors used by MAC. The FAC stage consists of several SME matchers working in parallel, and set in literal similarity mode. Each SME matcher starts by creating all possible correspondences between identical relations, and then between the arguments of those relations. The program then incrementally coalesces local matches into larger mappings that meet the constraints of *parallel connectivity* (if two predicates are mapped, their arguments must also be mapped) and *one-to-one mapping* (elements in one analog must map to at most one element in the other analog). Finally, FAC estimates the quality of global mappings as a function of their size, their depth, and the semantic similarity of their corresponding objects. As with the selection component of the MAC stage, FAC chooses the base items whose comparison with the target yields the highest score, plus all other items within the 10% of such score. The main differences between the literal similarity mode and the analogy mode reviewed in Chap. 1 can be summarized as follows:

1. When SME runs in analogical mode the system neither computes attribute matches nor builds correspondences with them unless dictated by relational mappings. As reviewed in Chap. 1 when analyzing the heat-flow/water-flow analogy, SME disregarded the similarity between the base element *water* and the target entity *coffee*—despite both being liquid and sharing other features—on the grounds that they were not integrated into a structural interpretation of the comparison. When running in literal similarity mode, however, SME would create matches between these objects and their attributes, since they could potentially participate in an eventual interpretation of the comparison, for example, in terms of a surface match wherein higher-order relational mappings cannot be obtained. To illustrate, consider comparing the water-flow scenario with a retrieved situation in which a vial is filled with water to a higher level than a tube. As there is no connection between the vial and the tube, this later system does not maintain any relevant relational commonality with the water-flow target. But as both situations still maintain similar entities (vial, tube, and water) and an isolated lower-level relation (more water in the vial than in the vase) the comparison represents a surface match. While SME's analogy mode would not be sensible to this type of resemblance, its literal similarity mode would detect it.

2. When running in analogy mode, SME's structural evaluations disregard resemblances between entities. In the heat-flow/water-flow analogy presented in Chap. 1, the entities paired within the winning interpretation were not alike. But what if the water-flow target were now compared to a retrieved base analog in which two solutions of different concentrations are separated by a semipermeable membrane? As a consequence of a difference in osmotic pressures, water is diffused across the membrane, and into the less concentrated solution. When running in analogy mode, SME would regard this analogy between water flow and solution flow as having the same quality as that between water flow and heat flow. In contrast, the literal similarity mode would show a preference for the water-flow/solution-flow analogy on the grounds that in both situations a difference of *pressure* causes a transfer of *matter*.

Forbus et al. (1995) conducted a series of simulations by way of feeding MAC/FAC's LTM with different databases. In one of the simulations, MAC/FAC's MLP consisted of the base stories from the nine sets used in Gentner et al. (1993), along with the first-order relations (FOR) matches of those sets (see Sect. 2.2), which served as distractors. Using either the literal similes, the mere-appearance matches, or the analogous base stories as target situations, retrieval rates were 100%, 88%, and 56%, respectively, thus reproducing the ordering observed in behavioral studies. In another simulation using the literal similes, the mere-appearance matches, the analogous matches, and the FOR matches of the abovementioned sets as LTM items, the target stories of those sets led to retrieval patterns that reproduced both the ordering and the absolute retrieval rates observed in behavioral studies: literal similes (78%) > mere-appearance matches (44%) > analogous matches (11%) > FOR matches (0%). Table 2.3 shows the contribution of both the MAC and the FAC stages to the final output of MAC/FAC.

The mean number of memory items produced by MAC is 3.3 and the mean number of memory items finally accepted by FAC is 1.5. As expected, FAC succeeds in its job as a structural filter on the MAC matches: It accepts all of the literal similes proposed by MAC and some of the partial matches (the mere-appearance, analogous, and FOR matches), while rejecting most of the unrelated stories that belonged to different story sets. It might seem surprising that FAC outputs more mere-appearance than analogous matches, when it normally would prefer analogous over

**Table 2.3** Proportion of different match types retrieved by MAC/FAC (from Gentner et al., 1993)

| | MAC | FAC |
|---|---|---|
| Literal simile | 0.78 | 0.78 |
| Surface match | 0.67 | 0.44 |
| Analog | 0.33 | 0.11 |
| FOR match | 0.22 | 0 |

*Note.* Memory contains 36 stories (the literal, mere-appearance, analogous, and FOR match for each of nine story sets). The probes were the nine base stories mere-appearance matches. The reason is that it receives twice as many mere-appearance than analogous matches from the MAC stage. For a comparison between MAC/FAC and the computational models proposed by the multiconstraint theory (ARCS and LISA), see Thagard et al. (1990) and Hummel and Holyoak (1997).

Albeit based on different representational and processing assumptions (e.g., while MAC/FAC relies on symbolic representations, LISA relies on distributed representations), dominant computational models of analogical retrieval present a reasonably good fit to the behavioral results arising from the problem-solving and story-reminding traditions. However, it could be argued that despite the fit to empirical data, the adequacy of a computer model as a psychological theory ultimately relies on the validity of the empirical data on which they are grounded. Hence the question: are experimental results representative of analogical retrieval *at large*? Chap. 3 will review a more recent generation of naturalistic studies scrutinizing the production of analogies as it takes place outside of laboratory settings.

## References

Anoli, L., Antonietti, A., Crisafulli, L., & Cantoia, M. (2001). Accessing source information in analogical problem-solving. *Quarterly Journal of Experimental Psychology, 54*, 237–261.

Barnett, S. M., & Ceci, S. J. (2002). When and where do we apply what we learn? A taxonomy for far transfer. *Journal of Experimental Psychology: General, 128*, 612–637.

Day, S. B., & Goldstone, R. L. (2012). The import of knowledge export: Connecting findings and theories of transfer of learning. *Educational Psychologist, 47*, 153–176.

Duncker, K. (1945). On problem solving. *Psychological Monographs, 58*, 270.

Forbus, K., Gentner, D., & Law, K. (1995). MAC/FAC: A model of similarity-based retrieval. *Cognitive Science, 19*, 141–204.

Gentner, D., Rattermann, M. J., & Forbus, K. D. (1993). The roles of similarity in transfer: Separating retrievability from inferential soundness. *Cognitive Psychology, 25*, 431–467.

Gick, M. L., & Holyoak, K. J. (1980). Analogical problem solving. *Cognitive Psychology, 12*, 306–355.

Hummel, J. E., & Holyoak, K. J. (1997). Distributed representations of structure: A theory of analogical access and mapping. *Psychological Review, 104*, 427–466.

Keane, M. T. (1987). On retrieving analogues when solving problems. *Quarterly Journal of Experimental Psychology, 39*, 29–41.

Newell, A., & Simon, H. A. (1972). *Human problem solving*. Englewood Cliffs, NJ: Prentice-Hall.

Raynal, L., Clement, E., & Sander, E. (2017). Challenging the superficial similarities superiority account for analogical retrieval. In G. Gunzelmann, A. Howes, T. Tenbrink, & E. Davelaar (Eds.), *Proceedings of the 39th annual conference of the cognitive science society* (pp. 2957–2962). Austin, TX: Cognitive Science Society.

Reed, S. K., Ernst, G. W., & Banerji, R. (1974). The role of analogy in transfer between similar problem states. *Cognitive Psychology, 6*, 436–450.

Ripoll, T. (1998). Why this makes me think of that. *Thinking and Reasoning, 4*, 15–43.

Spencer, R. M., & Weisberg, R. W. (1986). Context-dependent effects on analogical transfer. *Memory & Cognition, 14*, 442–449.

Thagard, P., Holyoak, K., Nelson, G., & Gochfeld, D. (1990). Analog retrieval by constraint satisfaction. *Artificial Intelligence, 46*, 259–310.

Wharton, C. M., Holyoak, K. J., & Lange, T. E. (1996). Remote analogical reminding. *Memory & Cognition, 24*, 629–643.

# Chapter 3
# The Naturalistic Tradition

The main strength of laboratory studies resides in their stringent level of experimental control. The artificial manipulation of variables, ideally coupled with random assignment of participants to different conditions, allows the experimenter to separate the effects of a variable under study from those of an undefined number of extraneous factors, both known and undevised. The results obtained through these techniques are replicable: any person reproducing the original experimental conditions should obtain similar results, albeit with the degree of certainty that inferential statistics provide.

Despite the virtues of the experimental approach to psychology research, several authors have pointed to their pitfalls and limitations (e.g., Brunswick, 1943; Lewin, 1943), especially when it comes to generalizing laboratory results to real-life situations. As an example, Brunswick (1943) has argued that the representativeness of the conditions inherent in an experimental setting (e.g., experimental tasks, stimuli, context, etc.) are as or more important that the representativeness of the sample of participants. Even though Brunswick did not live long enough to witness the impact of his claim, the concern with the ecological validity in psychological research flourished during the decade of 1970, stretching to current times. At the onset of the cognitive study of analogical retrieval, however, the design of experiments and the preoccupation about ecological validity followed separate tracks. For example, while it could be argued that the profuse overutilization of the military and the tumor stories in studies of analogical problem-solving surely serves to align and compare the results of a new study with most of the studies that predate it, Devil's advocates could contend that it is far from clear that the results obtained with those stimuli are representative of problem-solving in general, let alone spontaneous analogizing at large. The studies reviewed in this chapter had the potential of taking the empirical literature on analogical retrieval a long way in the quest of generalizing the results of the experimental tradition (see, e.g., Chap. 2) to a more diverse and naturalistic range of conditions.



M. Trench, R. A. Minervino, *Distant Connections: The Memory Basis of Creative Analogy*, SpringerBriefs in Psychology,
https://doi.org/10.1007/978-3-030-52545-3_3

## 3.1 Analogies Produced by Scientists

An empirical tradition of high face validity concerns self-reports of analogy use by expert scientists. For example, Johannes Kepler left written record of the numerous analogies that allegedly helped him account for the inversely proportional relation between a planet's distance to the Sun and the speed of its translation trajectory. As reviewed in Chap. 1, a noteworthy analogy extensively analyzed by Kepler was that between the propagation of a hypothetical motive force emanated by the Sun and the propagation of light emanated from a candle: just like the density of light radiation, which decays at an inverse square function of its distance from the source in which it originated, so does the circular motion through which the Sun propels the planets of the solar system, thus causing the outer planets to orbit at comparatively lower speeds. Similarly, Nersessian (1992) analyzed the role of analogies and models in the creative breakthroughs of Faraday and Maxwell. In certain cases, an in Wertheimer's (1945) classic analysis of the development of Einstein's special relativity or Thagard's (1999) analysis of the recent discovery of the bacterial origin of stomach ulcers, the review of scientists' notebooks and scholarly exchanges can be further enriched by retrospective interviews with the scientists.

The temporal resolution of these investigations depends on the sources available to the researcher. While rather coarse when only relying on publications, it can get much finer—say, on a scale of days—to the extent that laboratory notebooks and correspondence are available. However, historical records do not allow for minute-by-minute sequences of the scientists' thoughts (Klahr, 1999). Worse yet, the data are subject to all of the self-reporting biases that make retrospective verbal reports less accurate than verbal protocols (Ericsson & Simon, 1993). As pointed out by Nersessian (1992), every piece of the recorded data is itself a reconstruction by its author. But how do scientists *really* reason? As opposed to retrospective studies, the naturalistic observation of scientists could potentially overcome both the coarse temporal resolution and the subjective nature of the written record available to historical approaches.

By means of protocol analyses inspired in the work of Ericsson and Simon (1993), Kevin Dunbar embarked in the laborious task of recording the exchanges maintained by the members of four prestigious laboratories of molecular biology during a whole year. On top of the sources of information that are typical of historical investigation (publications, retrospective interviews, laboratory notebooks, grant proposals, etc.), he recorded the weekly meetings of each team, held interviews with individual scientists prior and after the meetings, and analyzed informal conversations maintained by lab members when they considered that an interesting result was about to take place. The biologists observed by Dunbar (1997) used analogies quite often. An analysis of the 99 analogies collected during 16 plenary meetings revealed that most of them served one of three different purposes: generating hypotheses, designing and perfecting experiments, and providing illustrative explanations to other members of the group. Upon classifying these analogies into three levels of semantic distance (same organism, different organism, and extra-biological), the authors were able to analyze the influence of these goals on the types of analogies produced. When the goals were either to design/fix experiments or provide illustrative explanations, within and between-organisms analogies were used equally frequently. When facing unexpected results, the semantic distance seemed to augment as the sequence of unexpected results got increasingly longer. However, only in two occasions the analogies reached outside the realm of biology. With regard to the analogies directed to generate hypotheses—arguably the activity most strongly associated with scientific discovery—3 were drawn to the target organism and 20 to other organisms. The fact that none of the analogies connected different domains suggests that, even among experts, distant analogies are rarely produced. When they appear, they seem more directed to communicate an idea to others than to genuinely suggest novel causal accounts for a given phenomenon. When biologists were asked to reconstruct the path that led to a significant discovery some time after the discovery was made, they often forgot the sequences of short-distance analogies that very incrementally led to the discovery, and often replaced them by more appealing ones that did not show up in the observational record of their work. These results led Dunbar to suggest that the historical record on the use of distant analogizing in science might be somewhat biased, granting a more central role to analogy in discovery than it really deserved. However, the highly technical and idiosyncratic nature of microbiological phenomena might make it particularly difficult to draw sensible analogies to other domains. Hence, the observed patterns might not be predicted to generalize to the use of analogies in other fields of inquiry.

In a more recent observational study, Chan, Paletz, and Schunn (2012) analyzed the conversations held by two interdisciplinary groups of research scientists while interpreting data sent by two Mars rovers during their initial exploration of Martian surface in search for evidence of present or past existence of water. Having landed

**Fig. 3.1** A humorous representation of Kevin Dunbar's naturalistic observations of molecular biologists at work

almost simultaneously on opposite sides of Mars, each of the rovers sent the data to its corresponding research team in the Jet Propulsion Laboratory. Both teams reviewed their data somewhat independently from the other team (for those who are curious, let us just say that the mission was a success). The transcripts of 11 h of discussions revealed 94 analogies, with 70% serving problem-solving purposes and 30% serving other purposes, such as communication. With regard to their semantic distance, 72% were classified as "near analogies" due to involving either the same mission (e.g., noting that a current rock has features and/or possibilities and challenges similar to the rock analyzed by the scientists the week before), or pertaining to the domains of geology and/or atmospheric science. Far analogies (e.g., when a scientist compared a rover instrument planning problem with a chess game) were those that mapped to sources outside of the Mars Exploration Rover mission or, more broadly, to disciplines different from geology/atmospheric science. By crossing the presence of analogies in a block of utterances with the presence of indices of uncertainty (e.g., words like "maybe," "I believe," "perhaps," etc.), the authors observed that cumulative uncertainty predicted the generation of analogies, and that analogies served to reduce levels of psychological uncertainty.

As opposed to molecular biology and geology, Psychology is a discipline whose unobservable theoretical phenomena foster the construction of analogical models. From Plato's cave analogy to Freud's hydraulic analogies or the computer metaphor embraced by cognitive scientists, the study of mental processes has yielded an important corpus of thematically distant analogies. Using a methodological approach similar to that of Dunbar (1997), Saner and Schunn (1999) concentrated on Dunbar's hypothesis that long-distance analogies arise once a discovery was made, and mainly to illustrate an idea to others. As opposed to lab meetings, which correspond to an early stage of the discovery process, the information presented in colloquia tend to correspond to later stages of scientific inquiry, at which conclusions and theories are typically well developed. These extended presentations targeted experts as well as less specialized audiences, and tended to compile highlights from a sequence of studies instead of excruciating details about a single study. Hence, the authors reasoned that the group/colloquium distinction would capture much of the life cycle from ongoing discoveries to final presentations of discoveries.

Saner and Schunn (1999) compared the analogies produced by psychological researchers during five laboratory meetings of between 70 and 90 min with those presented by three invited scholars during a series of colloquia organized by the Department to which the research groups belonged. Colloquia also ranged from 70 to 90 min in length, and topics included cognitive psychology, social psychology, and cognitive neuropsychology. Of the total of 67 analogies produced, 37 came from the lab transcripts and 30 from the colloquia. An analysis of the goals of the analogies revealed that the majority of analogies were merely *mentioned* (i.e., employed to illustrate ideas) instead of actually *used* to subserve reasoning, with a higher proportion of just mentioned analogies in colloquia than in lab meetings. More critically to our purposes, only 20% of the analogies were drawn to extra-psychological phenomena (e.g., when explaining the consumption of psychological resources in terms of the brain's consumption of glucose). About 80% of the analogies were drawn to other psychological topics, irrespective of setting (e.g., when comparing a current study to a previous psychological study). Crossing these variables revealed an interesting pattern: within-domain analogies were just as likely to be used for reasoning in colloquia as in lab meetings. By contrast, between-domain analogies were very likely to subserve reasoning in colloquia, but never within lab groups. Taken collectively, the on-line observation of scientists seemed to demonstrate a continuity between the results of psychological experiments on analogical retrieval reviewed in Chap. 2 and those arising from expert scientists, thus suggesting that the predominance of distant analogies in the historical record of scientific discovery could be dangerously biased.

## 3.2 Analogies Produced by Journalists and Politicians

Khong (1992) presented a detailed historical study on the role of analogies during the political discussions that preceded the US' invasion to Vietnam. The study featured extensive interviews with the protagonists of a heated debate that had taken place behind closed doors. Khong reconstructed the exchange of analogies between the Vietnam situation in the first years of the 60s and other military episodes that took place shortly before. Those who related the intrusion of north Vietnam's leader Ho Chi Minh into South Vietnam with either Hitler's invasion to Poland or North Korea's invasion to South Korea anticipated the success of an overt military intervention. In contrast, a minority of analysts compared a hypothetical invasion by the USA with France's failure to reassert their prewar colonial domination over South Vietnam. Those who believed in this later analogy, anticipated insurmountable difficulties, similar to those suffered by France during a long agony that stretched from 1946 to the definitive defeat in the village of Dien Bien Phu, in 1954. Even though Khong's investigation was not intended to answer questions about the nature of memory but rather about decision making in politics, it shows evidence that the analogies entertained by political decision makers were drawn to very similar episodes. But as it is often the case in political matters, the winds would suddenly change.

Towards the end of the century, the population of the Canadian province of Quebec held a heated debate about the perspectives of becoming independent from Canada, which led to a referendum on October 30, 1995. Electors had the choice of voting "yes" for becoming a new country, or "no" for staying in Canada. Both sides campaigned extensively, and the "no" side won by the slimmest of majorities (51%). By then affiliated to McGill University, Isabelle Blanchette and Kevin Dunbar took advantage of this opportunity to compile and analyze the analogies that appeared in three important newspapers from Montreal during the final week prior to the referendum (Blanchette & Dunbar, 2001). In their own words, the aim of their investigation was "to determine whether the types of constraints that have been identified in the cognitive laboratory also govern naturalistic uses of analogy and whether analogy use in naturalistic contexts suggests further constraints on the mechanisms underlying analogy use" (p. 730).

From a total of 234 analogies that appeared on +400 articles, around three-quarters were drawn to situations outside the fields of economy or politics. Chosen domains were as diverse as agriculture, family, sports, magic, and religion. For example, a representative analogy from the pro-independence side was "Québecers don't want to feel at home in the rest of Canada, what they want is to build their own home." Just like in the above example, the great majority (85%) of the source analogs were found to be emotionally moving, therefore transferring their emotional valence to the target domain. The emotional valence of the analogies was not related to the side, but rather to the goal of the analogizer: 80% of the sources used to support one's own position had a positive emotional valence, while 60% of sources used in analogies meant to attack the opposing position had a negative connotation. The goal of the analogy, however, did not have an effect on the range or on the source categories used, which were mostly interdomain.

The contrast between these results and those arising from previous naturalistic studies was interpreted by the authors in terms of the pragmatic orientation of the analogies. While the experts scrutinized by Dunbar (1997) or Khong (1992) were addressing other experts, the journalists and politicians involved in the independence referendum were addressing the general population. As the general population probably lacked domain knowledge about the fate of other similar episodes, the analogies selected by the journalists had to move away from the target domain (for a more detailed exposition of this argument, see Olguín, Trench, & Minervino, 2017).

When analyzing their results, Blanchette and Dunbar (2001) acknowledged the limitations of their methodology for inspecting the cognitive processes that led to the employment of the surveyed analogies. On the one hand, it is unclear whether the influence of the audience on the type of analogies takes place during or after the retrieval phase. It could be the case that journalists were mostly reminded of very similar cases, but later discarded them in favor of more distant analogies, judged to be more comprehensible and persuasive. To assess this possibility, Isabelle Blanchette analyzed the analogies presented by journalists in verbal vs. written communications. The lack of differences between these settings led her to conclude that the prevalence of distant analogies in written communications could not have originated in a conscious selection. Consequently, the authors went on to interpret their results as indicating that when people are drawing on their own memories to build analogies for a realistic and meaningful target situation, the retrieval of analogical sources is less constrained by superficial similarity than suggested by the experimental tradition.

While this is indeed a plausible hypothesis, Holyoak and Thagard (1995) had advanced a somewhat more parsimonious account for the eventual predominance of distant analogies in both science and politics. They argue that the range of the analogies might reflect the amount of knowledge that has been accrued at any given moment in time. At early stages of a discipline, as when politicians strived to concern the population about the dangers of communism or Kepler strived to understand the relation between the speed of a planet and its distance from the sun, the lack of a fully developed explanation allowed for the use of distant analogies (e.g., appealing to the *domino effect* to conceptualize the potential consequences of any communist annexation), of poor predictive power but clear potential for illustration. When a wealth of domain knowledge has accumulated, near analogies are both available and preferred, since they better support theory formation and prediction.

## 3.3 Analogies Generated by Design Engineers

Following Dunbar's (1997) in vivo methodology, Christensen and Schunn (2007) carried out a systematic observation of the analogies generated by a team of 19 professional design engineers from an internationally recognized company, as they generated new products in the domain of medical plastics. During the weekly product development meetings, which lasted between 30 min and 2 h, common activities included brainstorming, developing concepts, solving design problems, planning data collection, sketching and evaluating mock-ups and prototypes, conducting experiments, as well as discussing and exchanging knowledge about end users and production methods. The transcripts of the meetings that took place during the 5-month duration of the study revealed a mean of 11.3 analogies per hour of verbal data, showing that analogies were frequently employed during product development meetings. Analogies were coded as within-domain when they remained within the domain of medical plastics, as in existing products from the same company or of competitors. Analogies were coded as between-domain when they were drawn to domains outside of medical plastics as, for example, to the auto industry, biology, sports, etc. In line with the results of Dunbar (1997) with molecular biologists, the analogies used by design engineers while evaluating or anticipating problems of a new design were mainly within-domain. In line with the results of Blanchette and Dunbar (2001), however, superficially dissimilar analogies were more frequently used than close analogies when the goal was to communicate or explain ideas to other members of the group. Examples of between-domain sources included potato print, credit cards, zippers, children's slides, shoes, Christmas decorations, Venetian blinds, and lingerie. But quite surprisingly, distant analogies were almost as frequent as superficially similar analogies when the goal was to solve a problem. As in Blanchette and Dunbar's (2001) off-line study of the analogies produced by journalists and politicians, Christensen and Schunn's on-line observation of design engineers was able to document a profuse use of distant analogies during the core productive activity of experts in a given field.

## 3.4 Analogies Proposed by Managers and Economists

Bearman, Ball, and Ormerod (2007) studied the analogies produced by expert managers while analyzing cases as a part of workshops following the "Teaching with Cases" methodology. These experts were all post-doctoral academics employed in management schools within universities, and thus had a wealth of advanced domain knowledge, in some cases accrued over 10 years of experience. They worked in groups of four participants during their target case analyses, and received 2 or 3 cases to discuss, depending on the group to which they were assigned. Participants analyzed the cases in two different settings: initial, small group discussions which lasted between 30 min and 2 h, and subsequent large group discussions lasting for approximately 1 h. The aim of the case analyses was for groups to design a package of recommendations that could be implemented by one of the organizations described within the case. The expert groups produced 34 analogies in 6 h of discussion, a mean rate of 5.6 analogies/h. Only one of the observed analogies involved similarity at only the relational level, without there additionally being a degree of superficial similarity. Thus the majority of analogies that were made by participants were near analogies.

Kretz and Krawczyk (2014) analyzed the analogies produced by faculty and graduate students of Economy during weekly reading group meetings at the University of Texas. The authors documented an average of 16 analogies per hour of discussion. In sharp contrast with Bearman et al.'s (2007) professional managers, graduate-level economists produced an even proportion of superficially similar and superficially dissimilar analogies. An analysis of the relation between the purpose of the analogy and its semantic distance revealed that participants tended to use distant analogs when trying to generate visual images, and close analogs for the purpose of exemplification.

## 3.5 Analogies Developed by Educators

In another observational study, Richland, Holyoak, and Stigler (2004) analyzed the spontaneous use of instructional analogies that appeared in 25 US eighth-grade mathematics classrooms, videotaped as part of the Third International Mathematics and Science Study (TIMSS; Stigler, Gallimore, & Hiebert, 2000; Stigler, Gonzales, Kawanaka, Knoll, & Serrano, 1999). A total of 103 analogies were recorded. Each lesson had a mean of 4.1 analogies, with a range of 1 to 11. Sources of analogical comparisons were coded either as an outside-math phenomenon (15%), a schema, or general rule with no numbers or non-math context (12%), a decontextualized math problem, i.e., involving only numbers (40%), a contextualized math problem, i.e., set on a non-math context (19%), or multiple sources for a given target (16%). As in prior observational studies of experts, the degree of surface similarity between the sources and the targets varied greatly as a function of the type of knowledge the teacher was trying to impart. While instruction of procedures was grounded mainly on near analogies, (e.g., showing how to reduce $15xy^2z^4/25x^3y$ by pretending it simply consisted in 16/20), the transmission of conceptual knowledge tended to be rooted in superficially dissimilar or even non-mathematical comparisons (e.g., explaining the difference between a circle and a circumference by mapping both concepts, respectively, to an orange vs. the peel that would remain if the interior of the orange were completely sucked out). In the words of the authors, the fact that most of the analogies involved minimal perceptual similarity between source and target objects was taken to suggest that "in the classroom setting, teachers are more successful than typical laboratory participants at developing structural analogies" (Richland et al., 2004, p. 49).

To summarize, a growing body of naturalistic studies on the spontaneous use of analogy in real-world activities reveals a lot of variation in the number and proportion of far analogies that are produced. Far analogies are primarily associated with the activity of communicating ideas to others, but have been observed to subserve the core productive activity of experts in some fields of expertise. Hence, professionals of various fields seem to flexibly access superficially dissimilar sources from their background knowledge in a manner that is certainly not predicted on the basis of traditional experimental results.

## 3.6 Computer Models of Superficially Unconstrained Retrieval

The idea that people can retrieve past episodes through the use of structural cues was first articulated by Roger Schank in his theory of Dynamic Memory (Schank, 1982). Conceived as the end result of a series of computational developments that dealt with language comprehension (e.g., Schank & Abelson, 1977), dynamic memory theory posits that the same abstract structures that allow us to understand a given piece of text are responsible for its storage and later retrieval. When a new problem or situation is represented in working memory, it triggers the activation of structures at different levels of generality, which support the derivation of inferences and predictions related to its contents. In turn, these structures provide "labels" on the basis of which the represented situation gets indexed in memory, and on the basis of which they will get eventually retrieved every time a new situation happened to activate the same label. The assignment of labels and indexes to the contents of memory makes it unnecessary to exhaustively search all memory contents in a computationally costly way. The dynamic aspect of memory consists in this vast repertory of structures, as well as in the association of these structures to the items stored in memory. As this updating process takes place in slow but continuous fashion along extensive periods of time, it should therefore not compromise the functioning of the cognitive system at large.

In order to explain those circumstances where a target situation makes us be reminded of an analogous situation in another domain, Schank (1982) postulated the existence of abstract structures called *thematic organization points* (TOPs), potentially capable of indexing different situations in terms of the goals they pursue, and the factors that constrain the goal's achievement. As an example, both the *Military* story and the *Radiation* problem would elicit the top "destroy central target-means blockage." Even though indexing memory contents through structures such as TOPs would certainly simplify an otherwise prohibitive bulk of computation to take place at the time of need, the problem faced by humans and computers alike consists in analyzing the base and target situations to identify these structures. As has been noted by theorists both within and outside of the case-based reasoning tradition (e.g., Lange & Wharton, 1993; Owens, 1993) the extraction of these labels requires complex inferential processes, generally based on a wealth of domain-specific knowledge.

The MAC/FAC model reviewed in Chap. 2 was engineered to mimic human difficulties—rather than their facility—for interdomain retrieval. However, Gentner, Rattermann, and Forbus (1993) posit that as domain knowledge accumulates, base and target encodings will tend to contain a greater number of higher-order relations, thus promoting a uniform relational encoding via increasing the uniformity of the vectors' entries (the "indices" in case-based reasoning terms) and therefore the mutual accessibility of relationally similar situations within the domain. Hence, as domain expertise increases, "MAC/FAC's behavior may come to resemble that of a case-based reasoning model with some surface indexing" (Gentner et al., 1993, p. 567).

While the above approaches could account for purely relational retrievals within a domain, the growing number of observational studies reviewed in the present chapter suggest that professionals of different fields can frequently draw analogies to situations outside their domain of expertise, that is, to domains for which there is no reason to foresee encoding strategies more sophisticated that those laypeople would deploy. In order to simulate rapid relational retrieval without presupposing the availability of abstract encoding schemas like TOPs, O'Keefe and Costello (2008) developed a computational model that relies exclusively on similarities between structurally equivalent systems of relations. Based on the assertion that "contemporary theories of memory do not posit that memory is composed of distinct domains [the base analogs] as assumed in MAC/FAC" (O'Keefe & Costello, 2008, p. 5), the authors adopt a model of LTM that consists of a list of all instances of any given relation (regardless of the base analog in which they take part), which are in turn linked to the relations that were adjacent to them in the episodes that were encoded in LTM. As there is no explicit separation between the sources stored in memory, the model's current version can only deal with *connected representations*—that is, situations in which every relation is directly or indirectly linked to every other relation by means of one or more propositional links. Fed with this particular type of representations, O'Keefe and Costello's (2008) relational adjacency model performs retrieval and mapping in an intertwined fashion according to the following procedure. Given a target analog represented in higher-order relational form, the system starts by choosing an arbitrary relation (e.g., Relation 1), and selecting all instances of such relation in memory. Next, the system shifts its attention back to the probe, and after identifying any relation (or relations) that are adjacent to Relation 1 in the probe (e.g., Relations 2 and 3), it trims the initial cohort of instances of Relation 1 in LTM, preserving only those in which Relation 1 is adjacent to Relations 2 and 3. Using the same logic, the system continues checking to see if the relations adjacent to each instance of a relation in the probe are adjacent to the last level added to the partial matches in memory. When all the vertices in the probe have been visited, the search is complete and any complete matches will have been found. As a concrete example of its operation, consider the task of retrieving the representation of the solar system in response to having represented the structure of the atom in working memory according to the following propositional structure:

Cause [and (attracts (nucleus, electron), weight-difference (nucleus, electron)), revolve-around (electron, nucleus)].

To find a match to this probe, the system would start by choosing an arbitrary leaf relation from the probe (e.g., weight-difference) and selecting all the instances of such relation in memory. The list of instances (weight differences), which compiles all those instances, serves as an initial cohort of candidates for retrieval. Once this initial list is generated, the system looks for any relations that are adjacent to *weight-difference* in the probe (in this case, the relation *and*), and proceeds to check in which entries of instances (weight-difference) the relation weight-difference appears adjacent to the relation *and*, compiling these partial matches on the list instances (weight-difference; and), which consists of a subset of the original cohort. Using the same logic, the system goes back to the probe to see which of the relations that are different from *weight-difference* are adjacent to *and* (i.e., the relations *cause* and *attracts*), and uses this information to retain those entries in instances (weight-difference; and) that are in turn adjacent to both *cause* and *attracts*. This process is iterated until either the system finds a match containing all the relations from the probe in the correct structure, or fails to find such a match.

In its current state of development, O'Keefe and Costello's (2008) algorithm can only retrieve sources being completely isomorphic to the target. As a consequence of this limitation, the system is incapable of simulating any kind of analogical transfer, since transfer requires that elements in the source lack their counterparts in the initial representation of the target. Even though this state of affairs is certainly consistent with Blanchette and Dunbar's (2001) and Richland et al.'s (2004) finding that in natural settings analogizers use interdomain sources to communicate ideas to others (a situation where the reasoner starts-off with a complete representation of the target situation she wants to communicate), it falls short of accounting for Christensen and Schunn's (2007) finding that interdomain analogies are also frequent in situations like creative problem-solving, where an impoverished target gets illuminated by a source analog whose representation is comparatively more complete.

Albeit based on different representational and processing assumptions (e.g., while CBR models suppose the separation of episodes in LTM and the availability of schematic descriptors like TOPs or MOPs, O'Keefe and Costello's algorithm can operate without such structures) the above models are capable of simulating superficially unconstrained analogical retrieval. As stated at the end of the previous chapter, however, the adequacy of a computer model as a psychological theory ultimately relies on the validity of the empirical data on which it is grounded. Given that the profusion of purely relational retrievals observed in some observational studies would have not been predicted on the basis of the data obtained in experimental studies, the pressing question concerns which of both empirical traditions more accurately represents human ability for analogical retrieval. As mentioned at the end of Chap. 1, the answer to this question far exceeds the theoretical and computational modeling of similarity-based retrieval: it can potentially inform educational practices. If naturalistic studies are right, then either the problem of inert knowledge is an illusion, or else further research should carefully scrutinize natural environments in search for the Holy Grail of flexible transfer. If experimental studies are right, despite their alleged lack of ecological validity, then educator's concerns about the lack of transfer are granted, and theory-driven intuitions still represent a valid source of inspiration for the development of instructional interventions. Chapter 4 will review recent attempts to resolve this empirical inconsistency, in the hopes that an accurate diagnosis of our competence limitations will serve as a foundation for the eventual development of instructional interventions.

## References

Bearman, C. R., Ball, L. J., & Ormerod, T. C. (2007). The structure and function of spontaneous analogising in domain-based problem solving. *Thinking and Reasoning, 13*, 273–294.

Blanchette, I., & Dunbar, K. (2001). Analogy use in naturalistic settings: The influence of audience, emotion, and goals. *Memory & Cognition, 29*, 730–735.

Brunswick, E. (1943). Organismic achievement and environmental probability. *Psychological Review, 50*, 255–272.

Chan, J., Paletz, S. B., & Schunn, C. D. (2012). Analogy as a strategy for supporting complex problem solving under uncertainty. *Memory & Cognition, 40*, 1352–1365.

Christensen, B. T., & Schunn, C. D. (2007). The relationship of analogical distance to analogical function and pre-inventive structure: The case of engineering design. *Memory & Cognition, 35*, 29–38.

Dunbar, K. (1997). How scientists think: Online creativity and conceptual change in science. In T. B. Ward, S. M. Smith, & S. Vaid (Eds.), *Creative thought: An investigation on conceptual structures and processes* (pp. 461–493). Washington DC: APA Press.

Ericsson, K. A., & Simon, H. A. (1993). *Protocol analysis: Verbal reports as data* (2nd ed.). Cambridge, MA: MIT Press.

Gentner, D., Rattermann, M. J., & Forbus, K. D. (1993). The roles of similarity in transfer: Separating retrievability from inferential soundness. *Cognitive Psychology, 25*, 431–467.

Holyoak, K. J., & Thagard, P. R. (1995). *Mental leaps: Analogy in creative thought*. Cambridge, MA: The MIT Press.

Khong, Y. F. (1992). *Analogies at war: Korea, Munich, Dien Bien Phu, and the Vietnam decisions of 1965*. Princeton: Princeton University Press.

Klahr, D. (1999). *Exploring science: The cognition and development of discovery processes*. Cambridge, MA: MIT Press.

Kretz, D. R., & Krawczyk, D. C. (2014). Expert analogy use in a naturalistic setting. *Frontiers in Psychology, 5*, 1333.

Lange, T. & Wharton, C. (1993). Dynamic memories: Analysis of an integrated comprehension and episodic memory retrieval model. In *Proceedings of the thirteenth international conference on artificial intelligence* (pp. 208–213).

Lewin, K. (1943). Defining the field at a given time. *Psychological Review, 50*, 292–310.

Nersessian, N. J. (1992). How do scientists think? Capturing the dynamics of conceptual change in science. In R. N. Giere (Ed.), *Cognitive models of science* (pp. 5–22). Minneapolis, MN: University of Minnesota Press.

O'Keefe, D., & Costello, F. (2008). A fast computational model of analogical retrieval (and mapping). In B. C. Love, K. McRae, & V. M. Sloutsky (Eds.), *Proceedings of the 30th annual conference of the Cognitive Science Society* (pp. 2003–2008). Austin, TX: Cognitive Science Society.

Olguín, V., Trench, M., & Minervino, R. (2017). Attending to individual recipients' knowledge when Generating persuasive analogies. *Journal of Cognitive Psychology, 29*, 755–768.

Owens, C. (1993). Integrating feature extraction and memory search. *Machine Learning, 10*, 311–339.

Richland, L. E., Holyoak, K. J., & Stigler, J. W. (2004). Analogy use in eighth-grade mathematics classrooms. *Cognition and Instruction, 22*, 37–60.

Saner, L., & Schunn, C. D. (1999). Analogies out of the blue: When history seems to retell itself. In M. Hahn & S. Stoness (Eds.), *Proceedings of the 21st annual conference of the Cognitive Science Society* (pp. 619–624). Mahwah, NJ: Erlbaum.

Schank, R. C. (1982). *Dynamic memory*. Cambridge: Cambridge University Press.

Schank, R. C., & Abelson, R. P. (1977). *Scripts, plans, goals and understanding: An inquiry into human knowledge structures*. Hillsdale, NJ: Erlbaum.

Stigler, J. W., Gallimore, R., & Hiebert, J. (2000). Using video surveys to compare classrooms and teaching across cultures: Examples and lessons from the TIMSS video studies. *Educational Psychologist, 35*, 87–100.

Stigler, J. W., Gonzales, P., Kawanaka, T., Knoll, S., & Serrano, A. (1999). *The TIMSS videotape class-room study: Methods and findings from an exploratory research project on eighth-grade mathematics instruction in Germany, Japan, and the United States*. Washington, DC: U.S. Department of Education.

Thagard, P. R. (1999). *How scientists explain disease*. Princeton, NJ: Princeton University Press.

Wertheimer, M. (1945). *Productive thinking*. New York: Harper.

# Chapter 4
# Bridging the Divide Between the Experimental and the Naturalistic Traditions

> A subject walks into a psychological laboratory at 2:00 P.M. She is told a story about a problem and is given the solution to the problem. Then she is given a list of words to remember. Five minutes later (at 2:18 P.M.) she is given new problems to solve, one of which shares superficial similarity to the first problem and the second shares structural similarity with the first problem. She does not use the first problem to solve the second when the two problems only share structural similarity, but does use the first problem when it shares superficial similarity.
>
> A scientist is investigating the way that HIV works. She obtains a very strange result. To explain what happened she spontaneously draws an analogy to a genetic mechanism found in heat-resistant bacteria. A politician is trying to convince people to vote against independence in the Canadian province of Quebec. He says that voting for independence "would be like leaving an ocean liner for a lifeboat, without paddles, on a stormy sea."
>
> These three examples exemplify what I term the *analogical paradox*. Subjects in many psychology experiments tend to focus on superficial features when using analogy, whereas people in nonexperimental contexts, such as politicians and scientists, frequently use deeper more structural features. What is the cause of these different findings? How can we account for this apparent paradox?
>
> Dunbar (2001)

Long before the naturalistic evidence reviewed in Chap. 3 started to accumulate, Kevin Dunbar had famously called attention to the flagrant inconsistency between the results of the experimental and the naturalistic traditions: while the former studies point to insurmountable difficulties for retrieving distant analogs, the latter often reveal a proportion of distant analogs that could not be predicted on the basis of traditional results and theorization. He termed this inconsistency "the analogical paradox," and set forth to explain it.

## 4.1 Blanchette and Dunbar's Production Paradigm

An obvious mismatch between the experimental and the naturalistic traditions concerns the expertise of the participants with the target topic at stake: while the former invariably involves novice university students, the latter typically involves trained



M. Trench, R. A. Minervino, *Distant Connections: The Memory Basis of Creative Analogy*, SpringerBriefs in Psychology,
https://doi.org/10.1007/978-3-030-52545-3_4

experts. In line with prior experimental results showing that more knowledgeable students have greater chances of accessing distant analogs than their less knowledgeable counterparts (Novick, 1988), it could have been the case that the difference between naturalistic and experimental results mostly reflects participants' degree of expertise with the target issue. Another obvious mismatch between both traditions concerned the nature of the source analogs, rather than that of the targets. While in experimental studies participants have to retrieve a brief situation externally provided by the experimenters, in naturalistic studies the sources are taken from the broad and well articulated knowledge base of the reasoners, accrued along decades of experience.

In order to assess whether the unprecedented results obtained in observational studies of journalists and politicians would generalize to a non-expert population, Blanchette and Dunbar (2000) asked groups of college students (Experiment 1) as well as individual college students (Experiment 2) to generate persuasive analogies for another real-world political issue: the zero-deficit strategy. The zero-deficit strategy relates to the government's goal of eliminating the yearly differences between earnings and expenses that lead to national debt, a problem that all Canadian administrations had to deal with in the 1990s. Both the federal and the provincial governments had engendered an enormous deficit that had to be cut, and governments reasoned that they had to cut their spending to shorten the deficit. The authors explained to their participants that one view of the deficit problem was that the deficit must be cut no matter what the consequences, since interest runs on the accumulated debt and the government has to pay this interest or else it increases the amount of debt. This of course entails more expenses, therefore more borrowing, making the borrower increasingly more vulnerable and dependent upon the lenders. The other view agrees with the preoccupation about the ever-increasing public debt, but contends that deficit reduction cannot be achieved at any cost, since cuts to education, social welfare, and health services would translate into social chaos. Hence, other means of dealing with the deficit had to be envisioned. After receiving general information about the zero-deficit strategy for dealing with increasing debt, two groups of participants were asked to pretend they were hired by a non-profit organization as part of a nationwide campaign to sensitize both citizens and governments about the urgency (pro-zero deficit condition) or the dangers (anti-zero deficit condition) of a drastic cut to public spending. Upon being shown an example of the use of analogies in persuasion (George Bush's 1992 comparison between Saddam Hussein and Adolf Hitler to gain support for a military intervention in Kuwait), participants of both groups were asked to come up with as many analogies as they could to support and illustrate their corresponding position. Regardless of condition, in both experiments this *production paradigm* (as Blanchette and Dunbar termed it) elicited a high number of analogies. While in Experiment 1 each participant in a group either proposed or else participated in 5.5 analogies, individual participants in Experiment 2 proposed an impressive rate of 10.73 analogies during the 20 min of the task. More importantly, an analysis of the range of the analogies showed that two-thirds of the sources provided by the groups of Experiment 1, and 80.6% of those reported by the individual participants of Experiment 2 were drawn to semantic domains different from economics/politics and personal finances, such as illness/medicine (11.3%), animals/farming (4%), domestic tasks (2%), eating (5.6%), or natural resources/disasters (4.8%), among many others. The authors interpreted these results as providing an in vitro replication of the previous naturalistic observation that expert journalists and politicians can easily access long-distance sources while generating analogies to real-world target situations. Such results lend some support to the hypothesis that the main factor facilitating interdomain retrieval was related to the possibility of drawing on participants' own extra-experimental sources. However, the fact that the zero-deficit problem had also received some coverage in the media during the previous decade prevented the authors from discarding the possibility that the high rates of interdomain retrieval obtained with the production paradigm had originated in a higher familiarity with the target topic, as compared to that of most laboratory studies.

To assess this possibility more directly, Blanchette and Dunbar (2000) subjected some of the situations brought by participants of Experiments 1 and 2 to a two-phase paradigm (*reception paradigm*, in the authors terminology), similar to that of the story-reminding tradition (see Chap. 2). During a first phase, participants received four base analogs (two that could potentially support a pro-zero deficit position and two that could support an anti-zero deficit position), along with eight fillers, and had to rate them for pleasantness. After a brief distracter activity, they received two target situations (one in favor of zero-deficit and another one against it), plus two unrelated situations. As both targets pertained to political finances, the mere-appearance match of one of the pro zero-deficit target was surface-related to the anti-zero deficit target, and vice versa. Participants were asked to indicate to which of the situations read during the first phase each target situation reminded them of. They conceived this third study as a crucial experiment. If the results of their previous experiments had to do with the familiarity of the target topic, then the same pattern of distant analogizing should also arise under this more traditional procedure. If, on the contrary, the results had originated in the participants' possibility of basing their analogies on their own extra-experimental knowledge, then results should reveal the surface superiority bias typically obtained in experimental studies. In line with traditional experimental results, participants retrieved the superficially similar situation much more often than the purely structural analog, suggesting that participants ability to generate far analogies under a production paradigm was not an effect of their familiarity with the stimuli, but rather of the possibility of drawing on their own knowledge. Dunbar's (2001) own explanation to the analogical paradox posited that participants under a production paradigm are representative of the subjects in naturalistic studies in two important ways. As opposed to the reminding task used in the story-reminding studies, a meaningful and ecologically valid task such as generating persuasive analogies leads participants to extract the structural features of the target situation. On the other hand, the real-life situations employed as a basis of analogical arguments had been encoded in ways that emphasized their underlying structure, unlike the seemingly artificial stories that participants typically receive in both the problem-solving and the story-reminding traditions. Given that both the source and the target have been encoded with an emphasis on their structural properties, retrieval can be achieved based on shared structure irrespective of the amount of superficial overlap. In their own words, "retrieval of sources was not highly constrained by superficial similarity" (Dunbar, 2001, p. 116).

To gather further evidence for the power of the analogy generation task for eliciting a structural encoding, we carried out a study designed to compare the proportion of far vs. close analogies drawn from the extra-experimental memory of participants to the proportion retrieved from an experimentally presented set by the same group of participants, and during the same analogy generation task (Trench, Oberholzer, Adrover, & Minervino, 2009, Experiment 2). During the second phase of the experiment, participants received one of ten hypothetical scenarios wherein the main character is about to carry out an action with potentially negative consequences, and participants were asked to come up with analogies directed to dissuade the character from carrying out his intended action—i.e., a task virtually identical to that of Blanchette and Dunbar (2000). However, during a contextually separated previous phase participants had received several situations with the instruction to read them very carefully. One of them was either an intradomain or an interdomain analog to the hypothetical situation of the second phase. Given that the action performed by the character of this source analog engendered negative consequences, it was suitable for being used as a source situation for the analogy generation task. An analysis of the analogies drawn from participants' own extra-experimental memories replicated the majority of distant analogies found by Blanchette and Dunbar (2000). In contrast, an analysis of the experimentally presented situations reported by participants revealed a much more frequent retrieval of the close base analogs, as in experimental studies. Results seemed consistent with the idea that when participants are allowed to draw on their own extra-experimental memory in the service of persuasive analogy generation, the retrieval of participants' own memories seems less constrained by surface similarity than had previously been considered.

If naturally encountered situations are more likely to receive a structural encoding than the sources typically learned under experimental settings, where does such advantage originate? One salient mismatch between naturally encountered situations and typical experimental stimuli is that the former tend to include an auditory component, whereas the latter typically consist of written materials. Markman, Taylor, and Gentner (2007) demonstrated that an auditory presentation of both the base and the target analogs leads to better recall of purely relational matches than a written presentation. These results led them to speculate that such advantage could underlie the alleged advantage of naturally learned base analogs for interdomain retrieval. Another potentially important difference between naturally and experimentally encoded situations could reside in the fact that the former tend to be relatively more meaningful for participants (Hofstadter & Sander, 2013). Even though these hypotheses are appealing, the overarching claim that naturally acquired sources are advantageous for being retrieved in response to distant targets is, at minimum, counterintuitive.

### 4.1.1 *Reasons to Doubt the Superiority of Natural Encodings*

On the one hand, the temporal separation between learning and retrieval tends to be very brief in experimental studies and very large in real life, spanning from months to years, or even decades. On the other hand, both the physical and the functional contexts that surround learning and transfer tend to be weakly or moderately separated in the laboratory, and very large in natural settings. To complicate matters more, while the base and the target analogs typically used in experimental studies (e.g., the materials employed by Keane, 1987 or Gentner, Rattermann, & Forbus, 1993) tend to be cast in ways that maximize their structural overlap, the base and target situations that take part in naturalistic analogizing were not created with the explicit goal of giving rise to particular analogies, and thus present large differences in size, frequent structural mismatches, and a bulk of irrelevant information. Not surprisingly, many instructional interventions that have yielded positive results in the vacuum chamber of the psychological laboratory do not prove successful under the noisy conditions that characterize real educational environments. With these caveats in mind, we shifted from trying to explain the alleged facilitation of distant retrieval in naturalistic settings to trying to assess whether naturalistic settings in fact support this kind of superficially unconstrained retrieval.

### 4.1.2 *Reasons to Doubt the Adequacy of the Production Paradigm*

Even though Blanchette and Dunbar's (2000) production paradigm can exert a higher degree of control than purely observational approaches, it falls short of demonstrating a negligible role of superficial similarities during naturalistic retrieval. As compared to the two-phase procedure employed in experimental studies of analogical transfer, the production paradigm suffers from three important methodological shortcomings. The first one consists in failing to discern between true instances of analogical retrieval and cases of analogy fabrication (i.e., ad hoc invention of base analogs). To exemplify, once the analogizer has grasped the abstract structure of a target situation, she can trivially generate analogous cases by replacing a handful of target elements by other entities which could conceivable play similar roles (e.g., infections of diphtheria/cholera/choose your favorite infection/that were not treated in time, with the consequence that later treatment was more complicated than would have been otherwise). Second, even if some means of distinguishing true retrievals from fabricated cases were implemented, the proportions of near vs. far analogies reported by participants may not faithfully reflect the retrieval criteria followed by the cognitive system. Under particular pragmatics like persuasive argumentation, participants could avoid reporting several superficially similar sources, which will tend to be very similar to each other by virtue of their mutually shared surface features. Finally, even if some means of revealing all retrieved sources were devised, yet another shortcoming of the production paradigm would concern the unknown proportions of close vs. distant sources available in LTM. To illustrate, consider a participant retrieving 6 near and 6 distant sources during a given task. In case this participant had, say, 10 near and 10 distant sources in LTM, the retrieval probabilities of near and distant base analogs should not be regarded as being constrained by superficial similarity (60% of each type). If, on the contrary, she had 10 near and 60 distant sources in LTM, the same retrieval outcome would now indicate that the retrieval of base analogs was strongly constrained by superficial similarity (60% of superficially similar vs. 10% of superficially dissimilar sources). As this example illustrates, assessing the extent to which surface similarity determines the retrieval of naturally acquired base analogs requires knowing not only the number of far and near sources that got successfully retrieved, but also the number of both types of base analogs that were potentially available for being retrieved (i.e., retrieval probability = number of retrieved sources/number of available sources).

## 4.2 A Hybrid Paradigm Retaining the Best of Experimental and Naturalistic Studies

An exhaustive survey of all near and far analogs stored in a person's memory would of course be impossible to achieve. In order to circumvent this limitation, we took inspiration from the procedure first used by Chen, Mo, and Honomichl (2004) to

**Fig. 4.1** A humorous representation of the empirical paradigms used for investigating analogical retrieval. Left panel: Traditional 2-phase story-reminding paradigm, wherein experimenters place a structural and a superficial source in LTM and compare their probabilities of being retrieved. Right panel: Production paradigm, wherein the proportion of close vs. distant analogs may simply reflect their prevalence in LTM

assess analogical retrieval after longer intervals than two-phase experimental studies permit.

### *4.2.1 Early Studies on the Retrieval of Cultural Narratives*

In a series of experiments, Chen et al. (2004) assessed Chinese and American participants' retrieval of popular tales while solving analogous problems with different types of surface similarity. One of the popular tales known by a vast majority of Chinese participants had it that long ago in China, there lived a powerful Emperor who got an elephant as a gift from a ruler of a surrounding country. When the Emperor asked him what the weight of the elephant was, the ruler was embarrassed because even the biggest scale he had was too small to weigh the elephant. The story ends that the Emperor's younger son came up with the idea of putting the elephant on a boat, marking the new water level on the boat, then replacing the elephant by smaller stones until the water level reaches the same mark, and finally summing the weight of the smaller stones.

Upon signing an informed consent for participation on a study on problem-solving, participants received several problems to solve, only one of which was analogically related to the elephant story. By means of a factorial design, the target situations crossed presence vs. absence of similarity with the elephant tale at the level of target objects, with presence vs. absence of similarity regarding the available tools. Each target described a team of scientists on an exploration site who needed to figure out the weight of a large object (either an elephant or an asteroid) but did not have access to a scale large enough. After the presentation of the story, a set of illustrated items as potential elements for solving the problem was shown, including logs, boxes, rocks, trees, tools, rope, and buckets, that could be chosen in combination with a solution tool (either a boat or a spring platform) that could be used to weigh the goal object. After attempting to propose solutions, participants were directly queried about whether they were reminded of any story and/or event they had heard of or read before and, if so, what they were reminded of. Results revealed a surprisingly high overall percentage of participants being reminded of the elephant tale, with slightly higher rates by participants facing target problems with higher levels of superficial similarity. Despite the subtlety with which the superficial similarity was manipulated, even in the least superficially similar condition the target shared some identical relations and objects with the elephant tale (e.g., the goal of weighing a heavy object or the explicitly stated absence of a sufficiently large scale). In light of Keane's (1987) demonstration that even in experimental settings the presence of at least an identical or very similar object suffices to elicit retrieval (see Chap. 2), Chen et al.'s (2004) results are silent as to whether the popular tales would have been retrieved in the lack of such similarities.

In a related study, Dehghani, Gentner, Forbus, Ekhtiari, and Sachdeva (2009) investigated the retrieval of culturally shared fables during the resolution of dilemmatic choices between two alternative courses of action, an engaging and ecologically valid task somewhat akin to that of Blanchette and Dunbar's production paradigm. Like in Chen et al. (2004), even in the conditions with lower surface similarity the average retrieval rates were higher than those typically obtained in laboratory experiments. But also as in Chen et al., the fact that the target shares some identical elements with the culturally shared fable under study precludes using their results as a basis for inferring the extent to which naturalistic encodings and target tasks afford the retrieval of sources lacking surface similarities with the targets.

### 4.2.2 *Our Own Studies with Culturally Shared Base Analogs*

We reasoned that the idea of assessing the retrieval of culturally shared episodes could potentially help bridging the gap between the experimental and the naturalistic traditions on analogical retrieval, since it could easily be adapted in ways that retain the ecological validity of the production paradigm without sacrificing the methodological control of the standard, two-phase transfer experiments. Just as in the production paradigm, participants of Trench and Minervino (2015, Experiment 1) were asked to generate persuasive analogies for a realistic situation—an activity that according to Blanchette and Dunbar (2000) promotes an encoding of the targets that emphasizes their abstract structure. On the other hand, participants were asked to draw on their own extra-experimental memories, which according to Dunbar (2001) or O'Keefe and Costello (2008) receive a more abstract encoding than the source analogs typically used in traditional experiments. Having retained these distinctive advantages of the production paradigm, our procedure also preserved those features of the standard experimental paradigm that allow controlling the potentially distorting effects of analogy fabrication, report bias, and the eventually uneven availability of near vs. distant analogs in LTM. In order to control for this last factor, in Experiment 1 we assessed the retrieval of specific episodes from popular movies (the sources) during the task of generating persuasive analogies for realistic analogous situations that maintained different degrees of superficial similarity with such episodes (the targets). By way of restricting the data analysis to the retrieval of such specific base analogs, we were able to estimate (and therefore compare) the accessibility of close vs. distant naturalistic sources in terms of quotients between the number of retrieved cases and the number of cases potentially available for retrieval.[1] On the other hand, the improbability of fabricating a situation identical to a

[1] To some extent, the logic behind disregarding the bulk of analogies proposed by participants and simply concentrating the data analysis on the retrieval (or not) of this particular analog is somewhat reminiscent of techniques used in marine migration studies. As an example, consider the following questions: (1) Which is the likelihood that a mother whale that departed from California will make it to Alaska? and (2) How does it compare with the probabilities of a newborn whale? The success probabilities of mother whales can be compared to those of newborn whales without necessarily labeling all existing whales with satellite transmitters. It only requires taking the quo-

culturally shared episode ensured that all analogies built upon those episodes were originated in the retrieval of such items from LTM. Finally, the possibility of directly querying participants about whether they were reminded of the critical episode during the persuasive argumentation task was key to neutralizing an eventual bias in the report of near vs. far retrievals. As an example of our materials, we assessed the retrieval of "Jurassic Park" (a movie seen by almost all participants independently of the study) while dealing with an analogous situation that had either high or low levels of surface similarity with it. In "Jurassic Park," a millionaire has cloned dinosaurs from the Jurassic Period out of fossil DNA taken from a fossilized mosquito. Despite receiving expert advice about the difficulties involved in exerting total control over this kind of biological phenomena, the millionaire opens a park to exhibit the dinosaurs to the public, with the consequence that dinosaurs later break the security system of the park, and attack people. Superficially similar targets were derived by replacing base objects and relations with semantically similar ones. For instance, the superficially similar target of the "Jurassic Park" set stated that a businessman had replicated mammoths from the Pleistocene Era out of a frozen embryo preserved in a glacier, and was entertaining the idea of opening a zoo with mammoths on show. Participants' task consisted in employing analogies to dissuade the main character from pursuing the project, on the grounds that as the behavior of this kind of animals was not completely predictable, mammoths could potentially destroy the zoo cages, thus endangering people. In turn, superficially dissimilar target analogs were derived by means of replacing base objects and relations with new objects and relations that were semantically less similar than in the near targets. Keeping with the Jurassic Park set, the superficially dissimilar target stated that an astrophysicist was recreating Martian storms out of digital images captured by a space probe. The target ended up stating that the astrophysicist was planning to allow his colleagues to enter the experimental zone in order to study these storms. Participants had to propose analogies that could be used to dissuade the main character from pursuing his plan on the grounds that, as extraterrestrial meteorological phenomena are not sufficiently comprehended, they could potentially exert harmful effects on his colleagues.

Collapsing the results of all four sets of materials, superficially similar source analogs were retrieved in 70% of the trials. In contrast, superficially dissimilar sources were retrieved in only 15% of the cases. Figure 4.2 shows the contrast between close and distant retrievals for each of the four movies under study: *Jurassic Park*, *Shrek*, *Spiderman*, and *The Secret in Their Eyes*.

This strong effect of superficial similarities on the retrieval of naturally acquired base analogs was consistent with a long experimental tradition using artificial stimuli (see Chap. 2), and ran counter to the conclusions extracted from naturalistic studies (see Chap. 3). Given that the naturalistic encoding of the movies under study happened to have a strong auditory component, the observed effect of surface

---

tient between an arbitrary number of mother whales labeled in California and the subset of whales finally transmitting in Alaska, and comparing it to an analogous quotient corresponding to the newborn whales.

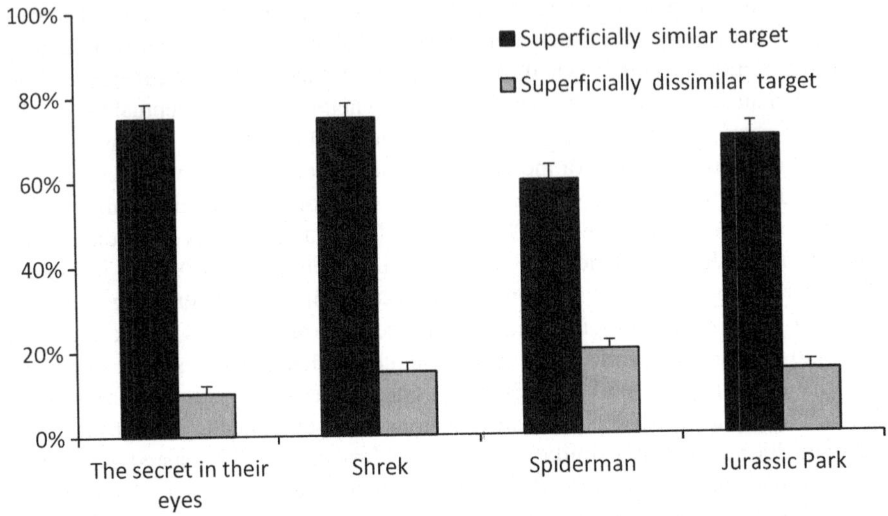


**Fig. 4.2** Retrieval rates of four popular movies during the processing of close vs. distant analogous situations (from Trench & Minervino, 2015, Experiment 1)

similarities also failed to support Markman et al.'s (2007) conjecture about the role of auditory presentation in rendering naturally encoded sources more retrievable.

Despite the robustness of our results—the high number of participants exposed to each of the sets allows considering each set as being itself a replication of the surface similarity effect— the fictional nature of the stories still prevented generalizing the results to the retrieval of naturally encoded situations at large. In an unpublished study with a similar procedure, we turned to the retrieval of another type of culturally shared episodes: extensively published political affairs. One of such episodes was the story of Sergio Schoklender, the older of two brothers who murdered their parents in the early 80s. During his confinement in prison, Schoklender's strong determination to pursue university-level studies as well as to resist harassment by other convicts called the attention of human rights activist Hebe de Bonafini, internationally recognized for having demanded an immediate accounting of all of the missing people, including her own sons'. Once released from prison, Schoklender became increasingly involved with the administration of Bonafini's foundation, and began to receive large amounts of government funding for the construction of welfare houses (the "shared dreams program"). By 2011, it became apparent that the foundation ran by Schoklender had defrauded the Argentine government, leaving numerous poor families without the houses that the government had funded. Participants in the superficial similarity condition were presented with a hypothetical situation involving a young man who had pursued in-jail studies of social care during his imprisonment for having poisoned his step-father. After reading that the director of an NGO was planning to hire him to supervise a community eatery for poor children, participants were asked to draw analogies that could be useful to dissuade the director of the NGO from hiring the prisoner, on the grounds that despite having paid his debt to society, his criminal background could still represent a threat to young children. In contrast, participants in the superficial dissimilarity condition were presented with a situation wherein a dog who had severely attacked its owners underwent a rehabilitation program, with the result that it learned to respect its trainers as well as people in general. After reading that the director of an institute for the blind was planning to involve the dog in an initiative to assist the blind in their daily activities, participants were asked to come up with analogies that could be used to dissuade the director from involving the dog, on the grounds that despite its impressive transformation, the aggressive background of the dog could still represent a threat to blind people. Just like in Trench and Minervino (2015, Experiment 1), participants easily recalled the critical culturally shared episode during the work with the superficially similar target, but failed to retrieve it in response to the superficially dissimilar scenario.

### 4.2.3 *Is There a Retrieval Advantage of Autobiographical Episodes?*

And what about the alleged advantage of personally significant episodes for being spontaneously recalled in the absence of surface similarities with the target? While it seems likely that the culturally shared episodes included in the above studies had been better learned than the artificial stimuli typically employed in experimental studies, they might have lacked the personal significance that characterizes autobiographical episodes. Unfortunately, the source–target paradigm has a serious defect that undermines the generality of the conclusions that experiments based upon it produce. According to Hofstadter and Sander (2013) when in real life we are in front of novel situations, the source episodes we retrieve spontaneously and effortlessly from our memories are, in general, extremely familiar. People depend on knowledge deeply rooted in their experiences over a lifetime, and this knowledge, which has been confirmed and reconfirmed over and over again, has also been generalized over time, allowing it to be transferred to all sorts of new situations. Taking into account Hofstadter and Sander's considerations about the key role of familiarity and personal significance during naturalistic analogical retrieval, in Trench and Minervino (2015, Experiment 2), we designed a procedure aimed at assessing the effect of superficial similarity on the retrieval of autobiographical episodes. During a first phase of the experiment, participants received descriptions of four different types of situations with the instruction to report known personal episodes conforming to any of those descriptions. While in half of the cases these schematic descriptions encompassed situations that would maintain high levels of surface similarity with the target analog to be presented in the following session, the other half encompassed situations that would be less superficially related to the future target. In one of these four sets of materials, the cue for collecting close sources asked for situations wherein the participant had made a quick fix to an artifact which proved partially successful, with the negative consequence that she ended up never seeking a fully adequate solution. In contrast, the cue used for collecting distant sources asked for situations wherein the participant had learned intuitively how to do something in a rather clumsy way, with the negative consequence that she ended up never learning the proper way of doing it.

Less than 2 weeks later, participants having reported an exemplar of at least one of the four cues were asked to participate in a study on analogical argumentation. This phase took place in a different location and was administered by different experimenters from those participating the previous in phase. After reading a realistic situation that, unbeknownst to participants, was either superficially similar or superficially dissimilar to the source they had reported during the previous phase, they were asked to draw analogies to real episodes they had experienced in order to dissuade the main character of that situation from carrying out his intended action. Following with the Quick Fix set used as an example, participants were asked to imagine that a friend of theirs had discovered that the WC was leaking, and that he was about to seal it temporarily with *suprabond* even though the result would not be completely satisfactory. Participants were asked to use analogies to known situations to dissuade him from fixing it by himself, since it could prevent him from ever seeking a completely satisfactory solution in the future. In line with the results of Trench and Minervino's (2015) previous experiment, an analysis of participants' responses revealed a strong and uniform effect of superficial similarities on the retrieval of the autobiographical episodes that participants had reported during the first phase of the experiment. These results failed to support the special status of familiar and mundane autobiographical episodes as candidates for superficially unconstrained analogical retrieval.

In a very recent study, Raynal, Clement, and Sander (2018) insisted on the special nature of autobiographical memories for supporting superficially unconstrained retrieval. They argued that the autobiographical episodes involved in our experiment may not be ideally suited for receiving an abstract encoding, since participants lacked pre-existing categories of events for these experiences. For example, even though the base analog of the Quick Fix set would belong to an eventual category of events wherein "a person solves a problem in an informal and suboptimal way, mistakenly believing that she or he will later hire an expert who will provide the appropriate and definitive solution," there is no reason to expect that upon a first encounter with an exemplar of this potential family of episodes, a person would naturally index it in such an abstract way. According to Hofstadter and Sander (2013), the abstract encoding of a familiar situation can give rise to a *conceptual skeleton* that can be reactivated in new analogous situations. This conceptual skeleton becomes stored in LTM in the form of a category of situations referring to more and more abstract features, as new analogous situations are encountered and confirm the structural core of this new category. As they become familiar concepts, these categories of situations should be spontaneously activated during the encoding of new analog situations, thus promoting an abstract encoding allegedly capable of supporting analogical retrieval. Based on these considerations, Raynal et al. set forth to assess whether distant retrieval would increase when participants can draw on their own autobiographical episodes in the service of target situations that could be readily classified in terms of a lexicalized relational category. To this end, they presented participants with stories like the following: "I had the idea to answer that I forgot my glasses when the photographer offered me to go to his exhibition. The reality is that I had my glasses with me but I did not want to go there" (an instance of *excuse*). Instead of asking participants to generate analogies—a task that they argued could induce participants to invent, rather than retrieve situations—they resorted to a "free-recall" instruction to report any personal memory that came to mind during the reading of the target. Near half of the analogs were drawn to domains different from that of the target, while only 12% were taken from the same domain. However, as we have extensively argued when analyzing the strengths and weaknesses of Blanchette and Dunbar's (2000) production paradigm, the obtained proportions may merely be the result of having more distant than near analogs in memory. How many instances of having invoked visual difficulties for not assisting to a cultural activity could Raynal et al.'s (2018) participants have had available in memory? Without doubt, many fewer cases than the sum of excuses coming from distant domains (e.g., adducing being stressed for not initiating a diet, adducing not feeling well for not assisting to a class, adducing being urged no to keep the house in order, and a long etcetera). As thoroughly discussed in Sect. 4.1.2, assessing the extent to which surface similarity determines the retrieval of naturally acquired base analogs requires knowing not only the number of far and near sources that were retrieved, but also the number of instances of both types of base analogs that were available for retrieval.

Olguín, Tavernini, Trench, and Minervino (2020, Experiment 2) reassessed the hypothesis investigated by Raynal et al. (2018), but controlling for the uneven availability of near vs. distant analogs in participants' memory. To this end, they adapted the procedure and materials followed by Trench and Minervino (2015, Experiment 2) so as to assess the retrieval of autobiographical episodes that had been categorized as instances of the same lexicalized relational categories that were later used to categorize the target. The procedure consisted in forming two groups of participants who, as determined by a questionnaire presented during a first experimental session, had experienced instances of a particular schema composed by two relational categories connected by a causal relationship (e.g., *bad behavior* at home caused *punishment* at home). During a temporally and contextually separated session, the experimenters presented the target analog as a case of one of the schemas for which participants had reported an instance during the prior session (e.g., bad behavior caused punishment), thus ensuring that the base and the target had been encoded in terms of the same lexicalized relational category. While half of participants received a target pertaining to the same domain as the reported episode (e.g., *home*), the other half received a target pertaining to a different domain (e.g., *school*). Participants were asked to provide one analogous story that had happened to them. Near analogs were retrieved in 27.27% of the cases during the second phase, whereas distant analogs were retrieved only in 4.55% of the cases. The retrieval superiority of close remindings generalizes the results of Trench and Minervino (2015, Experiment 2), providing further evidence that surface similarity influences retrieval even when autobiographical episodes are exemplars of lexicalized categories of events. The results from Olguín et al.'s study suggest that the uneven availability of close and distant analogs would be a reasonable interpretation of Raynal et al.'s (2018) results. It should be noted that even though the advantage of close over distant remindings obtained by Olguin et al. (2020) replicated that of Trench and Minervino (2015, Experiment 2), the absolute values where somewhat lower. One possible interpretation of this drop-off concerns the accessibility of competing sources by virtue of the familiarity of the relational categories employed: given a higher number of analogous cases, the restriction to report just one situation might lower the odds that participants will favor the critical source over an ocean of alternatives.

The low retrieval rates of distant autobiographical episodes is striking if one takes into account the fact that, just as in Raynal et al.'s (2018) study, such memories were likely encoded under conditions that should be considered to be extremely favorable for being retrieved after superficially dissimilar targets. That is, conditions wherein the source and the target had been categorized in terms of a known, and even lexicalized relational category. Counter to Hofstadter and Sander's (2013) and Raynal et al.'s (2018) claims, our results suggest that even though people might encode autobiographical memories based on abstract schemas like those involved in relational categories, they retain a bulk of specific detail that might constrain subsequent retrieval. To exemplify, think of a situation wherein a thief stole a watch that belonged to your father by way of threatening you with a bladed weapon. Will you discard the details of the robbery due to having encoded it as an instance of this category of events? Will this episode be just as likely to be activated after the steal of another watch with bladed weapons than after the steal of a personal video by means of a hack? While more research is certainly needed to clarify the role of superficial similarity during the retrieval of autobiographical episodes, the data collected so far does not seem to support the alleged advantage of these memories for supporting superficially unconstrained analogical retrieval.

Taken collectively, the results of the above experiments converge in demonstrating that superficial similarities play a crucial role in the retrieval of naturally encoded sources during analogy generation. They demonstrate that a superficial bias governs analogical retrieval not only in laboratory conditions but also during the retrieval of participants' own sources in the service of tasks of indubitable ecological validity.

## 4.3 Is There an Adaptive Advantage of Retrieving Distant Analogs?

The surface bias of our memory systems has intrigued researchers in analogy and similarity. As eloquently conveyed by Gentner et al. (1993, p. 567): "How can the human mind, at times so elegant and rigorous, be limited to this primitive retrieval mechanism?" In terms of adaptation, memory's tendency to base retrieval on readily processable surface cues can be thought to represent no big loss, since most things that look alike are alike relationally as well (the *kind world hypothesis,* Gentner, 1989; Medin & Ross, 1989). In the words of Dedre Gentner, "if something looks like a tiger, it probably *is* a tiger" (Gentner et al., 1993, p. 567). More critically, the environment in which our ancestors evolved was so dangerous that risk avoidance was the top priority. As an example, suppose that after experiencing an almost deadly encounter with an animal, a hominid later stumbled across another animal with similar (though not identical) visual appearance. Falsely assuming identicality with regards to both animals' behavior would surely incur some cost in terms of time an energy. But wrongly denying their identicality on the basis of surface mismatches could have a lethal outcome. As humorously depicted by Fig. 4.3, retrieving distant analogs would have likewise engendered dangerous situations, as they could distract a person from retrieving literally similar analogs signaling potential dangers.

Luckily enough, our everyday environment is not nearly as dangerous as that of our ancestors, such that we can dare disregard surface resemblances or be captivated by distant resemblances. However, there are a wealth of situations where retrieving literally similar sources still represents a better choice than retrieving superficially dissimilar analogs. As with category-based induction, where similar exemplars represent a more solid basis for inferences than dissimilar ones (Osherson, Smith, Wilkie, López, & Shafir, 1990; Rips, 1975), the fact that two situations maintain a wide array of perceivable resemblances increases the probability that less salient features will also be shared. As stated by Hofstadter and Sander (2013, p. 156), "every act of thinking, no matter how small, relies on such analogies, *and the tighter the analogy*, the more unavoidable the conclusions it leads to would seem to be" (italics added). To illustrate, imagine that you are to persuade somebody else that ingesting too much passion-fruit will make her stop liking it. Even though analogies with (a) somebody who got fed-up of eating too many lychees, and (b) somebody

**Fig. 4.3** Two different ways of missing a literal simile with potential implications for survival. Left panel: Disregarding surface similarities that could signal unobservable commonalities. Right panel: missing a literally similar situation due to being captivated by a remote association

**Fig. 4.4** A humorous representation of the dubious appropriateness of relying on distant analogizing during certain activities

who got tired of playing too many computer games, might probably obtain similar scores on traditional measures of analogical soundness, it seems clear that by relying on the same set of biological foundations, the precedent of having been fed-up with lychees confers greater probability to the conclusion.

The problem arises when a pressing situation—no matter if natural or artificial—is so novel or unique that literal similes are lacking. Despite the fact that superficially dissimilar analogs would now represent the most solid foundation for inductive inferences, the surface bias of the memory systems will favor the retrieval of shallow matches (Gentner et al., 1993), which involve similar elements but dissimilar structure. These idiosyncratic situations get increasingly common as we move into unfamiliar territories, and become ubiquitous in educational environments. What, if anything, can be done to make distant retrievals more likely? The second part of the present monograph will review promising avenues for facilitating access to knowledge that would otherwise remain inert.

## References

Blanchette, I., & Dunbar, K. (2000). How analogies are generated: The roles of structural and superficial similarity. *Memory & Cognition, 28*, 108–124.

Chen, Z., Mo, L., & Honomichl, R. (2004). Having the memory of an elephant: Long-term retrieval and use of analogues in problem solving. *Journal of Experimental Psychology: General, 133*, 415–433.

Dehghani, M., Gentner, D., Forbus, K., Ekhtiari, H., & Sachdeva, S. (2009). Analogy and moral decision making. In B. Kokinov, D. Gentner, & K. Holyoak (Eds.), *New frontiers in analogy research* (pp. 1–10). Sofia: NBU Press.

Dunbar, K. (2001). The analogical paradox: Why analogy is so easy in naturalistic settings, yet so difficult in the psychology laboratory? In D. Gentner, K. J. Holyoak, & B. Kokinov (Eds.), *The analogical mind: Perspectives from cognitive science* (pp. 313–334). Cambridge, MA: The MIT Press.

Gentner, D. (1989). The mechanisms of analogical transfer. In S. Vosniadou & A. Ortony (Eds.), *Similarity and analogical reasoning* (pp. 199–242). Cambridge: Cambridge University Press.

Gentner, D., Rattermann, M. J., & Forbus, K. D. (1993). The roles of similarity in transfer: Separating retrievability from inferential soundness. *Cognitive Psychology, 25*, 431–467.

Hofstadter, D. R., & Sander, E. (2013). *Surfaces and essences: Analogy as the fuel and fire of thinking*. New York: Basic Books.

Keane, M. T. (1987). On retrieving analogues when solving problems. *Quarterly Journal of Experimental Psychology, 39*, 29–41.

Markman, A. B., Taylor, E., & Gentner, D. (2007). Auditory presentation leads to better analogical retrieval than written presentation. *Psychonomic Bulletin & Review, 14*, 1101–1106.

Medin, D. L., & Ross, B. H. (1989). The specific character of abstract thought: Categorization, problem-solving, and induction. In R. J. Sternberg (Ed.), *Advances in the psychology of human intelligence* (Vol. 5, pp. 189–223). Hillsdale, NJ: Erlbaum.

Novick, L. R. (1988). Analogical transfer, problem similarity, and expertise. *Journal of Experimental Psychology: Learning, Memory, and Cognition, 14*, 510–520.

O'keefe, D., & Costello, F. (2008). A fast computational model of analogical retrieval (and mapping). In B. C. Love, K. McRae, & V. M. Sloutsky (Eds.), *Proceedings of the 30th annual conference of the cognitive science society* (pp. 2003–2008). Austin, TX: Cognitive Science Society.

Olguin, V., Trench, M., & Minervino, R. (2020). Retrieving a distant analog from memory in daily life is very unlikely, even in optimal conditions of encoding. In S. Denison, M. Mack, Y. Xu, & B. Armstrong (Eds.), *Proceedings of the 42 Annual Conference of the Cognitive Science Society* (pp. 2124–2130). Austin, TX: Cognitive Science Society.

Osherson, D. N., Smith, E. E., Wilkie, O., López, A., & Shafir, E. (1990). Category-based induction. *Psychological Review, 97*, 185–200.

Raynal, L., Clement, E., & Sander, E. (2018). Structural similarity superiority in a free-recall reminding paradigm. In C. Kalish, M. Rau, J. Zhu, & T. Rogers (Eds.), *Proceedings of the 40th annual conference of the cognitive science society* (pp. 2324–2329). Austin, TX: Cognitive Science Society.

Rips, L. (1975). Inductive judgments about natural categories. *Journal of Verbal Learning and Verbal Behavior, 14*, 665–681.

Trench, M., & Minervino, R. (2015). The role of surface similarity in analogical retrieval: Bridging the gap between the naturalistic and the experimental traditions. *Cognitive Science, 39*, 1292–1319.

Trench, M., Oberholzer, N., Adrover, J. F., & Minervino, R. (2009). La eficacia del paradigma de producción para promover la recuperación de análogos base interdominio. *Psykhe, 18*, 39–48.

# Part II
# Overcoming Competence Limitations for Retrieving Distant Analogs

# Chapter 5
# Interventions to Enhance the Initial Encoding of Source Analogs

A stringent test of knowledge acquitition concerns whether the learned information can be accessed whenever it becomes necessary for new, unforeseeable situations that arise in scholar, professional, or daily life. As we have seen, this seems especially challenging when the target problems do not maintain surface similarities with previous base experiences. In view of this, research has concentrated on developing techniques aimed at increasing the likelihood of retrieving analogous situations pertaining to distant domains.

The dominant trend in promoting far transfer has focused on how knowledge is initially encoded. It is reasonable to think that if people encode information well from the start, the probability of retrieving it when needed will increase. The contexts wherein the presenter of a topic might try to favor an appropriate encoding of the learning materials so as to render them more retrievable can vary to a great extent. Such contexts impose different constraints as to the time available for intervening, the nature of the tasks we can expect the learners to perform, or the extent to which we will be able to monitor the effects of our interventions. As an example, a class with just a few students would allow for intensive interventions in terms of the presented information, the elaborations asked to participants, and the possibility of receiving personalized feedback. In contrast, none of this is available to a journalist addressing a TV audience. The interventions to be reviewed in the present chapter will shift from the more intensive and demanding to those that are more lean and austere, but hence more widely applicable.



M. Trench, R. A. Minervino, *Distant Connections: The Memory Basis of Creative Analogy*, SpringerBriefs in Psychology,
https://doi.org/10.1007/978-3-030-52545-3_5

## 5.1 Constructing Schematic Representations of Source Analogs

It is well established that human memory search is guided by semantic retrieval cues (Baddeley, Eysenck, & Anderson, 2015). In line with this idea, any semantic aspect of a novel problem could potentially provide a link to a relevant analog (see Loewenstein, 2010, for a review). As it is obvious, there will be many cues to retrieve a problem from the domain to which the target pertains; for example, the radiation problem referred to in previous chapters is likely to call to mind prior knowledge about related medical problems. In contrast, a base analog from a disparate domain will lack such transparent resemblances. However, if a base analog is indexed as a case of a more abstract formulation, or schema, this schema may remain available in LTM such that a future encounter with an analogous situation might elicit the activation of said schema, thus facilitating a connection between the base and the target analogs (the "learning to encode principle," Medin & Ross, 1989). For example, an appropriate schema for the military and tumor situations would include the need of using an intense force to overcome a central target, plus the fact that employing said resource would endanger elements that need to be preserved. If the processing of the military problem elicited a schema like this, then there would be at least some probability that a similar schema will also arise during the later processing of the tumor problem.

But there is another way in which the encoding of source analogs in terms of more general formulations could pave the way for later retrieval. In those cases wherein the base and the target analog differ in surface features but maintain an identical system of relations, any intervention aimed at removing surface features from the base analog or lessen their importance might facilitate its retrieval when confronted with a distant target, since surface mismatches will no longer obscure a good structural match (Kurtz & Loewenstein, 2007).

### 5.1.1 *Fostering Schema Construction by Comparing Two Analogous Situations*

Gick and Holyoak (1980, 1983) laid the foundations for a longstanding tradition of studies aimed at assessing the effectiveness of strategies meant to promote a more abstract encoding of base analogs. Three components of Gick and Holyoak's original studies became the landmark of this tradition: (1) the basic structure of the two-phase paradigm, whose encoding phase allows for the introduction of diverse interventions to augment subsequent retrieval, (2) the employment of the military and the tumor problems, now converted in the "fruit fly" of the experimental approach to analogical problem-solving, and (3) the thesis that indexing the source analog under the umbrella of a more abstract schema that retained its essential aspects will render such analog more readily retrievable during the processing of distant targets.

As reviewed in Chap. 2, Gick and Holyoak (1980) had shown that the spontaneous retrieval of the military problem while confronting the tumor problem is very unlikely to occur. In Gick and Holyoak (1983), the fourth experiment set forth to investigate the induction of a problem schema arising from the comparison of two analogs, as well as its effects on transfer. The first part of the procedure was presented as an experiment on story comprehension; subjects had to produce summaries of each story as well as to write descriptions of the aspects they shared. Afterwards, participants had to propose solutions to the tumor problem. The central manipulation of Experiment 4 involved whether participants received a single base analog or received two structurally equivalent situations with the instruction to compare them. The second base analog was the story of *Red Adair*, wherein a large quantity of fire-retardant foam must reach the base of an oil well that was set on fire. There was no hose large enough to carry the quantity of foam needed to extinguish the fire. Using many small hoses at the same time, Red Adair was able to pump a large quantity of foam on the base of the well. While 45% of participants in the two-analog condition spontaneously generated the convergence solution, only 21% of participants in the single analog condition did so.

By analyzing the schemas produced by participants, Gick and Holyoak (1983) observed that only about 40% of participants described the similarities between the two analogs in terms of the "ideal" convergence schema. However, as the effect of comparison on transfer proved to be mediated by the quality of the produced schemas, Gick and Holyoak hypothesized that any manipulation facilitating schema formation would increase analogical transfer. In follow-up experiments, they found that coupling the comparison activity with either a verbal statement or a diagram highlighting the shared aspects of the sources had a clear positive effect on transfer. The verbal statement and the diagram improved the quality of the schemas, which once again proved instrumental to later transfer.

Catrambone and Holyoak (1989) replicated the results of Gick and Holyoak's (1983) Experiment 4, but including a relevant control condition overlooked by that previous study: the presentation of two base analogs, but without the instruction to compare them. Hence, they confirmed that the transfer advantage obtained by Gick and Holyoak had originated in the comparison task, and not in the mere exposure to an additional example. Another improvement introduced by Catrambone and Holyoak (1989) concerned the strength of the contextual separation between both phases. Brown, Kane, and Echols (Brown, Kane, & Echols, 1986; see also Brown & Kane, 1988) had shown that directing young children's attention to canonical structural components such as goals, obstacles, and solutions promoted spontaneous transfer. In a similar vein, Catrambone and Holyoak tested whether more oriented comparison instructions could induce participants to focus on the problem-solution structure of the analogs, therefore facilitating analogical transfer to a target problem across more stringent contextual separations. Two sets of comparison instructions were employed. The more directive one begun by framing the comparison in terms of the components of a problem-solution schema, then gave feedback about the schema constructed from the first two analogs, and finally introduced a third example to which participants had to apply the learned schema. Such conditions produced impressive rates of spontaneous transfer, even after delays as long as one week.

In more recent times, studies focusing on analogical transfer in domains as diverse as science (Kurtz, Miao, & Gentner, 2001), engineering (Gentner et al., 2016), negotiation (Gentner, Loewenstein, & Thompson, 2003), and anatomy (Kurtz & Gentner, 2013) give support to the position that carrying out comparisons between analogous sources facilitates the learning and transfer of relational content. Hence, the transfer advantage of source comparison represents one of the most robust interventions directed to alleviate our structural limitations for cross-domain transfer.

The interventions reviewed so far were designed to promote an abstract encoding of the source situation via comparing it to a structurally similar case. Other authors have sought to explore more austere ways of inducing participants to generate schematic representations, in the sense of avoiding the provision of a further example specifically tailored for the occasion. To this end, they concentrated on devising specific instructions and tasks on the basis of which learners could construct such schemas on their own.

### 5.1.2 Abstracting a Schema Out of a Single Example

Mandler and Orlich (1993) considered the possibility that different instructions could promote reformulations of a source situation at different levels of abstraction, which could impact later transfer to different extents. They speculated that the most convenient type of representation should be relatively abstract, so as to avoid irrelevant surface information about the analogs, but not overly abstract, so as to retain the essential aspects of the situation. In a first experiment, they manipulated the instructions given to participants for describing the military story. The *detail condition* was meant to promote a relatively concrete encoding of the military story: Participants were instructed to give a detailed description of the story, including specific characters, places, and actions. The *gist condition* was designed to provide the "optimal" schema of the story, but still in a concrete form: Participants in this condition were asked to briefly summarize the main points of the story by stating the General's goal, dilemma, and solution. In turn, the *abstract condition* focused on the relations among the goal, dilemma, and solution, rather than specific objects or actions: Participants in this condition were asked to abstract the principle underlying the story in a way that would reflect a more general solution. As the schemas produced by participants did not neatly correspond to the received instructions, participants were regrouped in terms of the schemas they generated. An analysis of participants' solutions to the tumor problem supported the idea that analogical transfer is affected by the way in which the base domain is encoded. All participants who described the base story at the abstract level transferred the base solution to the target. Participants who gave gist descriptions transferred more often than subjects who gave detailed descriptions, suggesting that the inclusion of concrete details of the base may not be entirely detrimental as long as the schema has been abstracted.

Because the number of participants who produced abstract descriptions was very small, a second experiment was conducted in which participants received training to ensure that they would describe the base story at the desired levels. Participants were assigned to one of four conditions: *no training* (control), *gist training*, *abstract training*, and *meta-level training*, a condition wherein instructions stimulated vapid uninformative descriptions that missed essential aspects of the base problem and/or its solution (e.g., "a person confronts a difficult problem but ends up coming up with a creative solution"). In the three experimental conditions, participants practiced with two unrelated situations and received continuous feedback from the experimenters. Once they had mastered the type of reformulation that corresponded to their experimental condition, they were asked to apply the learned strategy to the military story. After carrying out two distractor tasks, they had to solve the tumor problem. Participants in the control condition received the same treatment, but no training at all.

The control group showed a lower level of transfer than all of the experimental groups combined. A trend analysis of the three training conditions showed a curvilinear trend with increasing levels of abstraction, which peaked in the abstract condition. The study showed that transfer performance depends on the generation of abstract schemas capable of subsuming the base case. More crucially, they revealed that the most advantageous level for the purposes of later retrieval is one wherein surface content is reduced to a minimum, but without giving away any essential component of the source situation's gist. These results provided the first evidence that transfer performance could be enhanced without the provision of additional information about the sources, thus challenging the by then shared assumption that source comparison was unavoidable for successful schematization.

### 5.1.3 Generating Analogous Situations

Bernardo (2001) carried out a series of experiments aimed at determining if the activity of creating an analogous problem improves later transfer to distant targets. He hypothesized that such activity could lead students to explore the underlying structure of problems and to ground them in their own world knowledge. In this way, the structure could be better grasped and also become more meaningful. In Experiment 2, which was specifically concerned with retrieval, participants were asked to study a set of mathematical word problems of different types. The study problems were presented along with worked-out solutions, and a description of the relevant principles for solving the problem. Participants of the experimental group were asked to construct their own analogous problems after they studied each of the different problem types. The control group were simply asked to study the principles and worked-out solutions for the different problem types. After studying the given set of problem types, participants received a set of new analogous problems. For each of the new problems, they were asked to list any problem from the first set that they thought was similar to the new one. Participants in the experimental condition were more likely to recall the relevant problem than participants in the control condition. Results also showed that it did not make a difference whether the problem solver was successful in constructing analogous problems or not. In general, the study showed that the invention of analogous problems is a demanding and difficult task.

### *5.1.4 Shifting from Domain-Specific to Domain-General Language*

Clement, Mawby, and Giles (1994) reasoned that the presence of domain-general relations in the source and the target would be more advantageous than their domain-specific counterparts for retrieving the source analog during processing the target. In the last experiment of their study, experimental conditions differred in terms of whether the verbs used to convey analogous lower-order relations were more or less general (i.e., domain-specific verbs vs. more generic verbs). The base analog concerned the behavior of creatures called Plokes. After Plokes fight, the winner replaces his body parts damaged in battle with the undamaged ones of the loser. The target situation concerned the behavior of politicians: After politicians compete in a debate, the winner replaces his ideas that had been discredited in the debate with more successful ideas of the losing candidate. Whereas in the domain general version, the Plokes *contend* with one another in the battle, and the political candidates *compete* for Mayor, in the domain-specific version the Plokes *bite* and the political candidates *run* for Mayor.

After completing comprehension tasks applied to the source as well as to four distractors, participants received the target analog and were asked to decide if it was analogous to any of the first five passages. Consistent with predictions, the proportion of subjects accessing the base analog was higher among participants reading domain-general verbs than among subjects reading domain-specific verbs. Clement et al.'s (1994) intervention broke away from the idea that the abstract schemas had to be derived from the sources in order to promote transfer, therefore opening an avenue for interventions that operate on the way the base information is initially presented. This type of intervention would be most appropriate for those situations in which we cannot expect learners to elaborate on the source analogs, but we can nevertheless optimize the way in which such situations will be represented.

### *5.1.5 Shifting from Concrete to Idealized Representations*

Goldstone and Sakamoto (2003) examined whether there was an effect of training with concrete or idealized graphics on spontaneous transfer of a general principle known as *competitive specialization*. To grasp what motion parameters of individual ants govern the collective distribution of the ants across available patches, participants were left to experiment controlling commands such as, given a selected patch, the speed of the closest ant towards that patch or the speed of all other ants towards the same patch. The aim of the simulation was that participants discover that, in order for ants to end up evenly distributed across parches, the parameters should be set such that the closest ant should approach a selected patch somewhat faster than the rest. Participants were trained on the Ants and Food simulation either with concrete graphics (black ants and small fruit) or idealized elements (black dots and green blobs), as shown in Fig. 5.1. Afterwards, they were asked to explore a thematically remote instance of the competitive specialization principle in which, if the parameters are set in an appropriate manner, initially undifferentiated matrices progressively begin to differentiate from each other, so that each one begins to resemble a different category of letter inputs. Even though the concrete representation of the source domain led to a better comprehension of the dynamics that govern the food and ants scenario, as measured by a quiz, participants in the idealized condition showed better transfer to the target domain of Sensors and Inputs than in the concrete condition.

## 5.2 Assigning the Source Situations to Relational Categories

The traditional interest in how entity and artifact categories are learned and applied has recently sprung to the realm of relational categories, and in particular to categories of events, or *schema-governed categories* (Goldwater & Schalk, 2016). In contrast to members of entity categories, which share a set of probabilistic features and feature correlations, members of schema-governed categories such as *assassination* share a structure that can be instantiated by very different exemplars (Gentner & Kurtz, 2005; Goldwater, Markman, & Stilwell, 2011; Markman & Stilwell, 2001), such as "Fred thrust a knife into Gina's heart," "Mary had Bob drink poison," or "The offender disconnected the patient's oxygen supply." Schema-governed categories specify the structure of situations in terms of a network of semantic interdependencies that hold among the constituents of the concept in question. This network includes several variables and limits the types of entities or actions that may be bound to each of these variables. They also stipulate how the instances assigned to a variable constrain the instances that can be bound to other variables (Markman, 1999).

Many of the mundane situations that have previously been used as materials for studying analogical retrieval can be considered either instances of schema-governed categories or systems of schema-governed categories (e.g., in the famous Karla set from Gentner, Rattermann, & Forbus, 1993; an *aggression* results in a *deal*). It is in principle conceivable that the processes by which current exemplars of relational categories trigger the retrieval of further exemplars might differ from the mechanisms that establish memory connections between new and old situations that, just

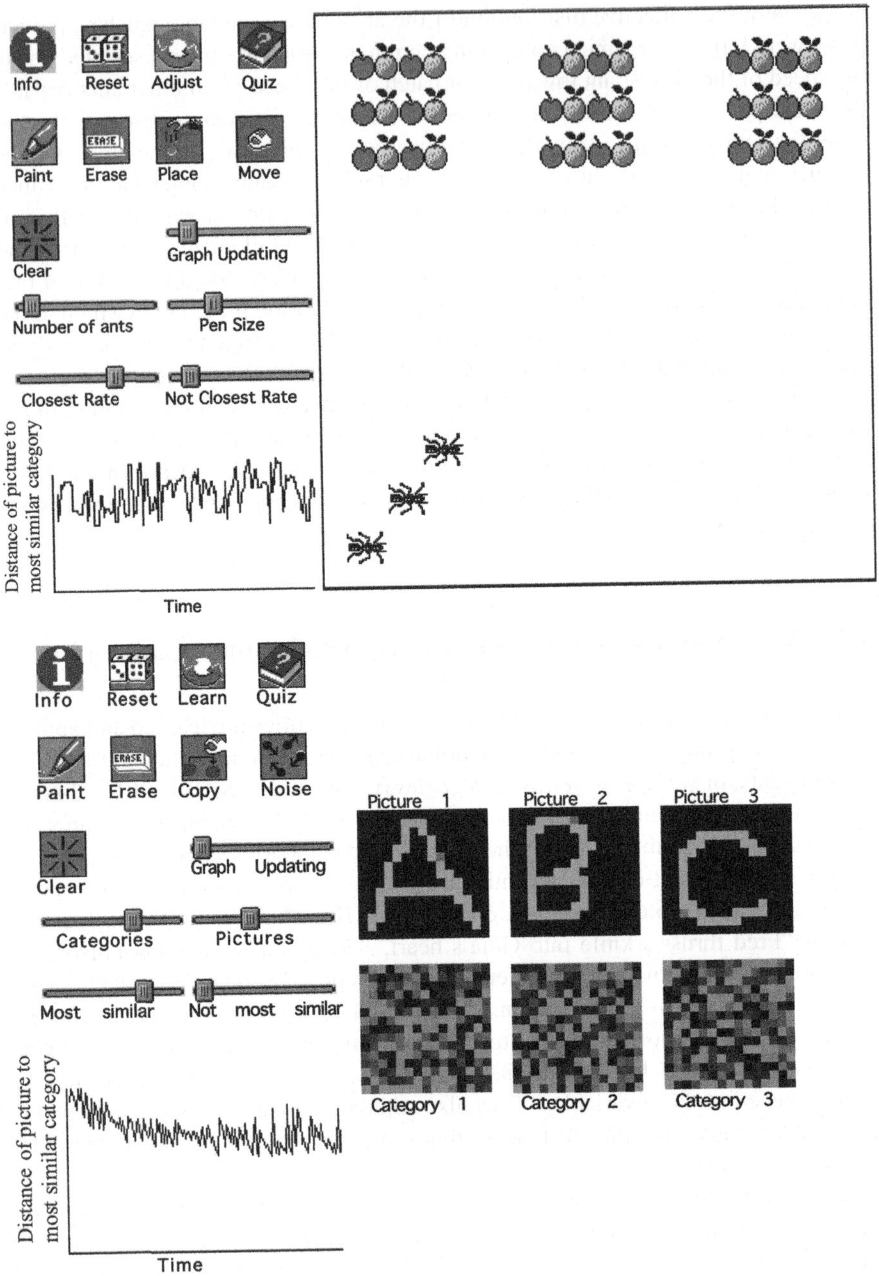


**Fig. 5.1** Snapshots of interactive environments corresponding to the Ants and Food (top panel) and the Sensors and Inputs (bottom panel) instantiation of the *competitive specialization* principle (from Goldstone & Sakamoto, 2003)

as the military and the tumor problems, do not constitute exemplars of this kind of categories.

A narrow set of empirical studies have begun to address what kind of similarities govern the process by which a target situation elicits the retrieval of further exemplars of the same category to which the target situation pertains. For example, Olguin, Tavernini, Pacella, and Minervino (2017) have shown that retrieval could be influenced by variables such as the extent to which the exemplars fare along critical dimensions of the category. As an example, if someone has stolen your mobile phone, you will more likely be recalled of a prior episode wherein your wallet was stolen, than of an episode in which you were stolen a chocolate, since the former is equivalent to the mobile phone episode in *importance*. Very recently, some studies have begun to address whether superficial similarities constrain how exemplars of relational categories are retrieved, as well as how to render their retrieval more likely.

### 5.2.1 Category Construction from Multiple Source Analogs

As reviewed along this chapter, constructing abstract representations of the source analogs has been regarded as the optimal way of increasing their likelihood of being applied to semantically dissimilar targets, on the grounds that the application of this same schema to a novel situation would aid in identifying the source situation among an ocean of competitors. In turn, the comparison of two or more sources has been considered the most powerful and straightforward way of eliciting the construction of these schemas by students. According to Kurtz and Honke (2019), however, the most efficient approach by which people make sense of a novel situation (e.g., a target problem) is to ask oneself to what category of situations the current case pertains. If this hypothesis were correct, then the most advantageous way of encoding a base analog would not be to index it as an specific instance of a long and complex abstract schema of the sort envisioned by Gick and Holyoak (1980) or Mandler and Orlich (1993), but rather as an exemplar of a category of events. The implications of this idea for the design of interventions to favor distant retrievals are rather straightforward: study tasks that promote viewing the base cases as instances of a category should be more effective than those that concentrate on abstracting a shared schema, but without treating it as a relational category.

Kurtz and Honke (2019) considered that one concrete way to promote viewing the base analogs as instances of a category could consist in having participants distill the principle shared by the instances via induction, and then having them apply such category to further cases. Participants had to discover the structure of a category via sorting a set of items into two different classes without receiving any feedback. In a first experiment, the materials to be classified were six text passages all superficially different from each other, but with half of cases pertaining to a relational category and the other half to a contrast category that maintained an alignable difference (Markman & Gentner, 1993) with the critical one (see Table 5.1 for a sample of materials). The authors compared the effectiveness of the category

**Table 5.1** Base and target passages used by Kurtz and Honke (2019)

| |
|---|
| Sample study example: Carnivorous plants eat insects. Insects do not naturally approach these plants; instead the insects are drawn to the scent of rotting animal flesh. Some carnivorous plants have the ability to emit chemicals into the air that mimic particular smells. How do the plants get their food? |
| Transfer test query: Once a virus enters your body, it infects a cell by injecting DNA into it. This viral DNA instructs the cell to produce thousands of copies of the virus. However, no matter how many virus copies are made the immune system always recognizes them as foreign and attacks when they exit the cell. How can the virus leave the cell AND survive? |

induction technique against the traditional approach of comparison-based schema abstraction. During the test phase, participants of both conditions were given an interdomain problem that could be solved by analogy to the source.

Participants with the category construction task were superior in retrieving base analogs (M = .56) than those receiving the standard comparison task (M = .35). A subsequent experiment replicated this central finding with a larger sample size. In brief, Kurtz and Honke (2019) introduced a new theoretical viewpoint suggesting that the problem of inert knowledge can be alleviated by means of viewing source analogs as members of a category instead of simply detecting a shared abstract pattern. The proposal is based on the intuition that categorization is a natural human activity when it comes to making sense of novel situations, one that can only succeed in eliciting the retrieval of source situations if the source situations have been indexed as exemplars of the category at stake. While their results only provide preliminary evidence in favor of this idea, they open a promising avenue towards new ways of conceiving analogical transfer.

### 5.2.2 *Highlighting Abstract Patterns Via Relational Labeling*

As we have seen, Clement et al. (1994) showed that if the base and the target include synonymous verbs describing corresponding base and target actions, the retrieval of superficially different analogs increases. In line with the thesis that the use of uniform relational language can improve the retrieval of distant analogs, Jamrozik and Gentner (2020) speculated that one factor that contributes to the uniform encoding that characterizes expertise in a field could be the availability and systematic use of vocabulary that captures the relational schema shared by the different situations (e.g., the use of words like *aggravated homicide* by lawyers). Jamrozik and Gentner advanced the hypothesis that applying a relational category label to an example stimulates representing the example in terms of a schema. Habitual use of the relational vocabulary of a domain can thus promote a uniform relational representation of the base and the target, and thereby increase the retrieval of distant analogs.

Jamrozik and Gentner (2020) had participants study one set of passages during an initial phase. After completing a filler task, participants received a new set of passages, and were asked to report to which of the situations of the preceding phase the current situations reminded them of. Each test passage had the same relational structure as one of the learned passages (e.g., *positive feedback loop*) (Table 5.2). In addition, to represent the competitive nature of retrieval, for each test passage there was another original passage from the same domain as the test passage. As we saw in Chap. 2, superficially similar non-analogs constitute a serious competitor for distant analogs. In a between-subjects design, the authors factorially manipulated the presence/absence of relational labels accompanying the source and the presence/absence of relational labels accompanying the target. For the label conditions, an additional sentence informed participants that the presented fact was a case of the corresponding relational category (e.g., "This is an example of feedback loop"). Results from the baseline condition revealed a majority of same-domain retrievals. However, providing relational labels during base encoding, target encoding, or during both instances improved the retrieval of distant analogs (Table 5.2).

## 5.3 Computer Simulations of Forward Transfer

Some models (including MAC/FAC, Forbus, Gentner, & Law, 1995; and LISA, Hummel & Holyoak, 1997) require that the total weight allotted to a memory representation be constant. If this is the case, the relations in a relational schema gain greater weight than those same relations would if embedded in a concrete-level exemplar, because the latter must also allot weight to its surface features. Thus, stronger relational matches can potentially account for the well-attested finding that relational schemas facilitate forward relational transfer. The earlier simulations in the area showed that target situations retrieve from memory, in order of decreasing likelihood: (a) literally similar cases, (b) mere-appearance matches, and (c) analogous cases. The key question is where relational schemas will fit in this ordering. As they lack the added similarity of matching surface features, schemas should be retrieved less often than literally similar cases. However, as relations receive greater weight in schemas than in cases, schemas should be retrieved more often than

**Table 5.2** Sample materials from Jamrozik and Gentner (2020)

| |
|---|
| Relational category: *Inoculation* |
| Source domain: *Psychology* |
| This is an example of inoculation. A person's attitudes can be changed by outside forces, such as persuasive advertisements. For example, after seeing a lot of convincing advertisements for cigarettes, a person's attitude toward smoking might become more positive. In order to prevent persuasive arguments from changing a person's attitude, the person can be asked to argue against weaker versions of these arguments. Later, if the person encounters more persuasive arguments, their attitude will not be affected |
| Target domain: *Medicine* |
| This is an example of inoculation. People can become infected by viruses that they encounter in their environment, such as influenza or chicken pox. For instance, someone can become infected through contact with a contaminated doorknob or appliance. To prevent infection, people can be exposed to a less virulent form of a virus, which their bodies will fight off by developing antibodies against it. If they are then exposed to the full-strength virus, their bodies can effectively fight off an infection |

analogous examples from memory. Thus, like mere-appearance matches, relational schemas should rank somewhere between literal similarity matches and analogous matches.

Gentner, Loewenstein, Thompson, and Forbus (2009) had MAC/FAC run on a subset of seven of the Karla the Hawk sets of stories (see Chap. 2). In the simulations, analogs had the same relational structure (both first-order and higher-order relations) but different entities and contextual features. Mere-appearance matches had the same surface features and first-order relations (specific actions and events) but different higher-order relational structure. Finally, first-order relational pairs matched in events but neither in higher-order relational structure nor in object attributes. To obtain abstracted schemas in an automated manner, they constructed two analogous stories having the same relational structure but different entities and contextual features, and used SME to carry out a structural alignment between them. The resulting schemas consisted of the common structure, including the consistent predicates that form the structure, the generic entities that play roles in the structure, and any consistent attributes shared by entities playing the same role.

In a first simulation, memory consisted of SME-generated relational schemas, analogies, mere-appearance matches, first-order matches, and unrelated examples. The MAC stage frequently advanced the schemas (57% retrieved) along with the mere-appearance matches (100%). In turn, FAC returned with equal probability relational schemas (57%) and mere-appearance matches (57%), with analogical sources being rarely retrieved (14% through MAC and FAC). Therefore, relational schemas were significantly better retrieved than the original, non-abstracted analogous matches (see Table 5.3). The most important finding in the first simulation is that analogical reminding is low, as traditionally obtained in empirical studies (14%), but becomes four times more frequent when a schema has been derived from the base analog.

As argued in more depth in Sect. 4.2.3, our memory systems have likely evolved to favor the retrieval of literally similar sources, of paramount import for both the obtention of valuable resources and the avoidance of threatening conditions. Hence, a subsequent simulation added literally similar items to the memory set in order to assess whether the obtained pattern would change. Literal similes were always retrieved (100%), thus reproducing the known empirical finding that literally similar examples are by great difference the most likely to be retrieved. The shallow matches (100%) and some of the relational schemas (43%) were retrieved by MAC,

**Table 5.3** MAC/FAC simulations of the retrieval advantage of abstracted sources (from Gentner et al., 2009, simulations 3 and 4)

| | Mere appearance | Analog | Literal simile | Schema |
|---|---|---|---|---|
| *First simulation* | | | | |
| MAC phase | 1 | 0.14 | | 0.57 |
| FAC phase | 0.57 | 0.14 | | 0.57 |
| *Second simulation* | | | | |
| MAC phase | 1 | 0 | 1 | 0.43 |
| FAC phase | 0.14 | 0 | 1 | 0.29 |

but not by FAC (14% and 29%, respectively), mainly because they were outscored by the literally similar examples, which match both in relational structure and in object attributes. Analogical matches were not retrieved at all, due to being outscored by the other kinds of matches.

Taken collectively, results from these simulations show that when a specific example is used as a probe, literally similar examples are by great difference the most likely to be retrieved, in line with Forbus et al.'s (1995) "primacy of the mundane" principle. But in those more pressing situations where there are no literally similar sources in LTM, relational schemas are as frequently retrieved as mere-appearance matches, albeit for different reasons.

In the above MAC/FAC simulations, the retrieval advantage of more abstract sources arises as a natural consequence of the way content vectors are normalized. Given that a fixed amount of weight is distributed among all the non-zero slots of the content vectors, the removal of object attributes entails allocating more weight to relational predicates, thus increasing the dot product between the now abstracted source and an analogous target. However, there is another possible psychological mechanism by which the abstraction of source analogs can increase their subsequent retrieval: If an abstract formulation of the source has been generated, such schema can remain available in LTM. If it happens to be reactivated during the processing of the target, then it can serve as a powerful cue to retrieve base analogs indexed as cases of such schema (Medin & Ross, 1989). Future empirical studies and simulations are needed to clarify the relative contribution of these two cognitive mechanisms to the retrieval advantage of abstracted sources.

## 5.4 Conclusions: Potential and Limitations of Source Elaborations

Despite the relative success of interventions aimed at promoting a more abstract encoding of the sources, a sensible question concerns whether improving people's initial encoding is the only way to favor the retrieval of analogs. As posited by Loewenstein (2010), if changing the initial encoding represents the only means for enhancing analogical reminding, then we can offer very little grounds for improvement to someone who has already acquired the relevant sources in a suboptimal manner. What is needed in these cases are interventions taking place at the time of retrieval.

## References

Baddeley, A., Eysenck, M. W., & Anderson, M. C. (2015). *Memory* (2nd ed.). New York: Psychology Press.

Bernardo, A. B. I. (2001). Analogical problem construction and transfer in mathematical problem solving. *Educational Psychology, 21*, 137–150.

Brown, A. L., & Kane, M. J. (1988). Preschool children can learn to transfer: Learning to learn and learning from example. *Cognitive Psychology, 20*, 493–523.
Brown, A. L., Kane, M. J., & Echols, C. H. (1986). Young children's mental models determine analogical problems with a common goal structure. *Cognitive Development, 1*, 103–121.
Catrambone, R., & Holyoak, K. J. (1989). Overcoming contextual limitations on problem-solving transfer. *Journal of Experimental Psychology: Learning, Memory, and Cognition, 15*, 1147–1156.
Clement, C., Mawby, R., & Giles, D. (1994). The effects of manifest relational similarity on analog retrieval. *Journal of Memory and Language, 33*, 396–420.
Forbus, K., Gentner, D., & Law, K. (1995). MAC/FAC: A model of similarity-based retrieval. *Cognitive Science, 19*, 141–204.
Gentner, D., & Kurtz, K. J. (2005). Relational categories. In W. K. Ahn, R. L. Goldstone, B. C. Love, A. B. Markman, & P. W. Wolff (Eds.), *Categorization inside and outside the laboratory: Essays in honor of Douglas L. Medin* (pp. 151–175). Washington, DC: American Psychological Association.
Gentner, D., Loewenstein, J., & Thompson, L. (2003). Learning and transfer: A general role for analogical encoding. *Journal of Educational Psychology, 95*(2), 393–408.
Gentner, D., Loewenstein, J., Thompson, L., & Forbus, K. (2009). Reviving inert knowledge: Analogical abstraction supports relational retrieval of past events. *Cognitive Science, 3*, 1343–1382.
Gentner, D., Levine, S. C., Dhillon, S., Ping, R., Bradley, C., Isaia, A., et al. (2016). Rapid learning in a children's museum via analogical comparison. *Cognitive Science, 40*(1), 224–240.
Gentner, D., Rattermann, M. J., & Forbus, K. D. (1993). The roles of similarity in transfer: Separating retrievability from inferential soundness. *Cognitive Psychology, 25*, 431–467.
Gick, M. L., & Holyoak, K. J. (1980). Analogical problem solving. *Cognitive Psychology, 12*, 306–355.
Gick, M. L., & Holyoak, K. J. (1983). Schema induction and analogical transfer. *Cognitive Psychology, 15*, 1–38.
Goldstone, R. L., & Sakamoto, Y. (2003). The transfer of abstract principles governing complex adaptive systems. *Cognitive Psychology, 46*, 414–466.
Goldwater, M. B., Markman, A. B., & Stilwell, C. H. (2011). The empirical case for role-governed categories. *Cognition, 118*(3), 359–376.
Goldwater, M. B., & Schalk, L. (2016). Relational categories as a bridge between cognitive and educational research. *Psychological Bulletin, 142*(7), 729–757.
Hummel, J. E., & Holyoak, K. J. (1997). Distributed representations of structure: A theory of analogical access and mapping. *Psychological Review, 104*, 427–466.
Jamrozik, A., & Gentner, D. (2020). Relational labeling unlocks inert knowledge. *Cognition, 196*, 104–146.
Kurtz, K. J., & Gentner, D. (2013). Detecting anomalous features in complex stimuli: The role of structured comparison. *Journal of Experimental Psychology: Applied, 19*(3), 219–232.
Kurtz, K. J., & Honke, G. (2019). Sorting out the problem of inert knowledge: Category construction to promote spontaneous transfer. *Journal of Experimental Psychology: Learning, Memory, and Cognition, 46*(5), 803–821.
Kurtz, K., & Loewenstein, J. (2007). Converging on a new role for analogy in problem solving and retrieval: When two problems are better than one. *Memory & Cognition, 35*, 334–341.
Kurtz, K. J., Miao, C., & Gentner, D. (2001). Learning by analogical bootstrapping. *Journal of the Learning Sciences, 10*(4), 417–446.
Loewenstein, J. (2010). How one's hook is baited matters for catching an analogy. In B. H. Ross (Ed.), *The psychology of learning and motivation: Advances in research and theory* (Vol. 53, pp. 149–182). San Diego, CA: Elsevier.
Mandler, J. M., & Orlich, F. (1993). Analogical transfer: The roles of schema abstraction and awareness. *Bulletin of the Psychonomic Society, 5*, 485–487.
Markman, A. B. (1999). *Knowledge representation*. Mahwah, NJ: Lawrence Erlbaum Associates.

Markman, A. B., & Gentner, D. (1993). Structural alignment during similarity comparisons. *Cognitive Psychology, 25*, 431–467.

Markman, A., & Stilwell, C. (2001). Role-governed categories. *Journal of Experimental and Theoretical Artificial Intelligence, 13*, 329–358.

Medin, D. L., & Ross, B. H. (1989). The specific character of abstract thought: Categorization, problem-solving, and induction. In R. J. Sternberg (Ed.), *Advances in the psychology of human intelligence* (Vol. 5, pp. 189–223). Hillsdale, NJ: Erlbaum.

Olguin, V., Tavernini, M., Pacella, L., & Minervino, R. (2017). Analogical retrieval mediated by everyday schema-governed categories. Paper presented at Analogy 2017: The Fourth International Conference on Analogy, Paris.

# Chapter 6
# Boosting Retrieval via Deliberate Search

> Ask, and it will be given you,
> seek, and you will find,
> knock, and it will be opened for you.
> Matthew 7:7

If a focus on surface similarity represented an adaptive means for ensuring the retrieval of past episodes in response to similar situations (see Chap. 4), it seems sensible to assume that the machinery in charge of this monitoring should continuously scrutinize LTM based on the contents of working memory. Consistent with this "automaticity" assumption, the great majority of behavioral studies were aimed at assessing the extent to which an experimentally provided source analog is *spontaneously* accessed during the processing of the target (e.g., Catrambone & Holyoak, 1989; Gick & Holyoak, 1980; Holyoak & Koh, 1987; Kurtz and Loewenstein, 2007; Spencer & Weisberg, 1986). Only a handful of studies (e.g., Blanchette & Dunbar, 2000; Keane, 1987; Trench, Oberholzer, & Minervino, 2009) had explicitly asked participants to think of analogous situations, and no studies had manipulated whether the deliberate intention to search LTM for analogous situations exerts an influence on retrieval outcomes. The main question that motivates the present chapter is thus as follows: Does the deliberate intention to search for analogous situations add anything over and above the automatic activation of similar cases in response to the contents of working memory? In case it did, under what circumstances?



M. Trench, R. A. Minervino, *Distant Connections: The Memory Basis of Creative Analogy*, SpringerBriefs in Psychology,
https://doi.org/10.1007/978-3-030-52545-3_6

## 6.1 Spontaneous vs. Voluntary Retrieval During Argumentation

Inspired in the profusion of interdomain analogizing obtained with a production paradigm, Trench, Olguín, and Minervino (2016) set forth to assess whether a deliberate attempt to search for analogous cases increases the number of distant retrievals as compared to the number of base analogs that would have been spontaneously retrieved. To this end, we confronted participants with a hypothetical scenario in which the protagonist was about to temporarily suspend the monthly payments of a loan due to other unanticipated expenses, with the idea of resuming payments at some point in the future. While participants in the *argumentation* condition were simply asked to generate arguments directed to dissuade the protagonist from carrying out his intended action on the grounds that later payments would be very hard to cancel, participants in the *analogical argumentation* condition were further asked to ground their persuasive arguments in analogous situations. In order to prevent participants of this condition from reporting base analogs not originated in retrieval processes, they were encouraged to base their analogies on past episodes which had happened to them or to others, or that were learned from verifiable sources such as newspapers, books, movies, etc. An analysis of participants' responses revealed that the deliberate intention to search for analogous cases can dramatically increase the number of analogies reported by participants (0.18 vs 0.73 analogies per participant). However, the fact that 62.1% of the proposals were intradomain revealed that the deliberate intention to look for analogous situations increases the number but not the type of the analogies that are obtained in the absence of such indication, which tended to be semantically similar to the target.

An additional condition helped us probe the power of voluntary search even further. Based on Gentner and Forbus' (1991) idea that our retrieval algorithms could conceivably be applied to subsets of LTM, we asked participants of a third condition to base their analogies on four domains different from economy: health, housekeeping, human relations, or the breeding of animals and plants. In contrast to the analogical argumentation condition, this last condition elicited the exact reverse pattern (65% interdomain vs. 35% intradomain), akin to the results of Blanchette and Dunbar (2000).

In an attempt to assess whether we could capitalize on voluntary search in ways that emancipate analogizers from the external provision of a set of domains, in a second experiment we compared the analogical argumentation condition against a further condition wherein participants were simply asked to base analogical arguments on situations outside the realm of economy. As expected, the proportion of distant analogies under this less directive intervention revealed a majority of interdomain analogies However, participants still proposed a non-negligible proportion of intradomain analogies (40%). How to interpret these results? While the enhanced employment of distant analogies suggested that participants were focusing search on either the distant domains suggested by the experimenters (Experiment 1) or on distant domains in general (Experiment 2), the fact that a non-negligible proportion of proposals were still intradomain suggests that the retrieval of these close analogies could not be successfully inhibited. Using an analogy from illumination, it seemed to us that rather than *focusing* search away from the target, participants were *broadening* the scope of the search, such that intradomain sources still fell within scope. But how to interpret the more basic finding that the mere indication to think of analogous situations tripled the number of analogies that were produced in the argumentation condition?

Given that prior research had shown that analogical arguments tend to be judged as being less persuasive than their factual counterparts (Keane & Bohan, 2004), one possible interpretation is that counter to the automaticity assumption shared by most theorists in the field, the argumentation task might not have sufficed to reliably initiate a search for analogous matches in LTM. On this account, the explicit prompt might have activated a search process that would not have been naturally triggered by the argumentation activity per se. However, another explanation for the difference between the argumentation and the analogical argumentation conditions could posit that the base analogs produced in response to the prompt to use analogies were not actually retrieved from LTM, but rather fabricated ad hoc (see Chap. 4). As an example of analogy fabrication, once the analogizer has induced a schema such as "if something that grows geometrically is not attacked early on, at some point it will grow at such a rate that you won't be able to reduce it," it is quite easy to envision the uncontrolled outburst of a tumor/infection/weeds/ants/(put your favorite plague) without having necessarily experienced it. On this second account, a prompt to think of analogous cases can boost the production of distant analogies by way of catalyzing the fabrication, rather than the retrieval of analogous episodes.

By means of embedding the argumentation task within a two-phase transfer paradigm less vulnerable to analogy fabrication, Amorín et al. (2017, Experiment 1) sought to assess whether the deliberate search for analogous cases can augment the chances of retrieving a distant source from LTM, as compared to a condition in which participants are not asked to base their persuasive arguments on analogous situations.

During the encoding phase, the base analog was incidentally presented to participants during a weekly class of a Psychology course. The base analog stated that at some time in the past, the USA and Russia had agreed to discontinue the production of the FLZ nuclear weapons, with extant missiles being kept in secure arsenals supervised by both countries. The story ended that after the mysterious disappearance of two missiles from one of the arsenals, a spread of nuclear radiation caused the death and illness of thousands of people (see complete materials in Table 6.1). During the transfer phase, which took place nearly 45 min after the presentation of the source story, participants were asked to participate in a study on argumentation. Both groups of participants received a target situation that maintained structural but not superficial similarities with the missiles story. It told that a dangerous human disease (the H flu) had just been eradicated, and that a group of scientists were planning to preserve the last samples of the virus for research purposes. Participants in the unprompted condition were asked to generate persuasive arguments that could be used to convince the scientists that the samples should be entirely destroyed.

**Table 6.1** Materials used in the argumentation and the analogical argumentation conditions, (from Amorin et al., 2017, Experiment 1)

| |
|---|
| *Base analog*. FLZ missiles are a nuclear weapon of enormous destructive power. After intense international negotiations, their production was discontinued in August, 1988. However, a considerable number of these weapons were stored in secured arsenals in Russia and the USA. Ten years later, after a mysterious disappearance of two warheads, an intense radiation was recorded, causing the death of hundredths of persons |
| *Target analog*. The H-Flu is a very contagious illness of potentially fatal consequences. It has recently been eradicated after a lengthy worldwide program of vaccination. The last samples of the virus are being kept at cryogenic state at the Vector Institute in Geneva (Switzerland). A group of scientists plans to keep them indefinitely for research purposes such as mapping its DNA or unveiling its mutation schemes |
| *[Analogical] Argumentation task*. [By using analogies to known situations] you need to convince the scientists that preserving the last samples of the virus can be potentially dangerous, and that they should therefore be destroyed. Please write down [analogical] arguments that could be used to dissuade the scientists from keeping the samples |

*Note:* The text between brackets was only included in the analogical argumentation condition

Participants in the prompted condition received a similar instruction, with the additional requirement to ground their arguments on analogies to known situations. While participants in the argumentation condition were spontaneously reminded of the base analog in only 7.81% of the cases, participants in the analogical argumentation condition were reminded of the base analog in 30.30% of the cases. Results thus replicated the benefits of voluntary search obtained by Trench et al. (2016), albeit through a procedure that was less vulnerable to the effects of analogy fabrication.

## 6.2 Spontaneous vs. Voluntary Retrieval During Problem-Solving

While no studies on problem-solving have manipulated whether participants are explicitly prompted to think of analogous situations during the processing of the target problem, across-studies comparisons suggest that participants' attempts to find a solution automatically elicit a search for base analogs in LTM. For instance, using roughly comparable materials, Keane (1987) and Gick and Holyoak (1980) assessed the retrieval of a base problem and its solution during a contextually separated problem-solving activity. Despite the fact that the former study (but not the latter) explicitly asked participants to look for analogous problems prior to attempting a solution, both obtained comparable rates of retrieval. Even though several procedural mismatches between the two studies should prevent extracting definitive conclusions, one can very tentatively explain the lack of differences across studies by positing that the activity of attempting to solve a problem might naturally trigger a search for analogous situations, such that an explicit prompt to remember analogous situations adds little over and above the mere disposition to solve the problem.

However, the fact that the retrieval of the base analogs was close to zero in both studies leaves open the possibility that the intrinsic difficulty for retrieving the military story during the processing of the radiation problem was so extreme that an otherwise successful prompt to look for analogous stories ended up not yielding noticeable effects. By explicitly manipulating whether participants are asked (or not) to think of analogous situations, in a recent study we aimed to reach beyond "across-studies" evidence by isolating the effect of voluntary search from other mismatching variables (Trench et al., 2020, Experiment 1).

The learning phase was administered by the instructor of an introductory Psychology course during a typical lecture, and was presented to participants of both groups as a reading-comprehension activity. Participants received three short stories of between 170 and 200 words and were instructed to read them very carefully, since they would have to answer questions about the stories without being able to reread them. While the first and last stories served as distractors, the central story was structurally similar to the target situation to be presented during the subsequent phase. In order to prevent an eventual advantage of voluntary search from being obscured by a floor effect, the military story used as base analog by Gick and Holyoak (1980) and Keane (1987) was replaced by an adapted version of the Fire Chief story (Gick & Holyoak, 1983), modified to be more structurally consistent with the tumor problem. This story told about a large fire at a botanical garden, which could only be suffocated by delivering several hundred liters of water per second. After being warned by the local guys that using his gigantic water pump to aspire water from a nearing pond would irreversibly damage the exotic specimens, the Fire Chief came up with the idea of recruiting several smaller pumps from neighboring fire divisions, and directing them to the fire from various locations. Participants were allotted 12 min to read the stories and 5 more min to answer two comprehension questions about each. In order to enforce a strong contextual separation, the transfer phase was administered by the experimenters between 30 and 45 min after the first phase, and presented to participants of both groups as a study on problem-solving. Participants in the analogical problem-solving condition received an instructional material on the effectiveness of solving problems by identifying analogous situations, which included two examples of how retrieving an analogous situation can aid in solving a current problem. These examples included an intradomain and an interdomain analogy so as to avoid biasing participants' conceptualization of ideal analogical comparisons in any particular direction. Participants were then presented with Duncker's (1945) radiation problem. The procedure and materials used in the problem-solving condition were similar to those of the analogical problem-solving condition, with the difference that participants received neither an instruction on the usefulness of analogy nor an indication to think of analogous situations prior to solving the problem.

Participants of both conditions were explicitly encouraged to include as many different solutions as they could think of within the 10 min frame allotted to completing the task. Since it is possible that some participants of either group had retrieved the base situation but did not derive a convergence solution to the target problem, a post-task questionnaire directly queried participants about whether any of the stories read during the text-comprehension activity had come to mind, even if just briefly, while they were generating solutions to the radiation problem. Participants responding "yes" to this question were further asked to state exactly which story (or stories) had come to mind during such activity. As in the previous experiment, the dependent measure consisted in whether participants reported having been reminded of the source story during the problem-solving activity.

In contrast with the manipulation of a voluntary search instruction during argumentation activities, participants in the analogical problem-solving condition were not reminded of the base analog more frequently than in the problem-solving condition (39.34% vs. 36.51%, $\chi^2$ (1, 124) = 0.11, $p$ = 0.7401, $\varphi$ = 0.03), a result that is consistent with our speculative across-studies comparison between the results of Gick and Holyoak (1980)[1] and Keane (1987).

The rate of analogical retrievals in our unprompted condition was far from a floor effect. Hence, the lack of differences obtained in the present study cannot be attributed to an extreme intrinsic difficulty for retrieving the source analog in response to the target situation. Rather, we interpret this null effect as indicating that the activity of attempting to solve a problem naturally elicits a conscious search for analogous situations in LTM, in light of which an explicit indication to conduct a memory search does not add much over and above this natural proclivity.

## 6.3 Spontaneous vs Voluntary Retrieval During Hypothesis-Generation

Dunbar's (1997) naturalistic observation of molecular biologists during their daily scientific and academic activities revealed that their analogies served three different goals: (a) solving a problem, (b) communicating an idea to others (akin to the persuasion task described in the previous section), and (c) generating explanatory hypotheses for puzzling phenomena. While problem-solving and effective communication have long been regarded as central targets of psycho-educational research, hypothesis generation has only recently come of age as a target of careful psychological investigation (Lombrozo, 2006).

Despite the seeming centrality of analogical reasoning in the genesis of scientific ideas (see Gentner & Jeziorski, 1989 or Sternberg, 1990, for reviews), no

[1] At first sight, our prompting manipulation might appear to be highly related to Gick and Holyoak's (1980) contrast between participants who solved the radiation problem before vs. after a hint to consider the stories read during a previous phase of their procedure. Despite this seeming commonality, the psychological constraints for capitalizing on such hint contrast sharply with those involved in our prompt to "think of analogous situations." As opposed to the latter case, in which the reasoner would need to probe the whole of LTM for potential matches, Gick and Holyoak's episodic reference to their learning set allows the reasoner to sequentially match the target against each of only three candidate situations, thus reducing an otherwise prohibitive computation to a much more manageable set.

experimental studies to date have delved beyond the information gained by retrospective (e.g., Gentner et al., 1997; Nersessian, 1992) or observational (e.g., Clement, 2008; Dunbar, 1997; Saner & Schunn, 1999) studies of the use of analogy by expert scientists. By applying a two-phase procedure to a hypothesis generation activity, Trench et al. (2020, Experiment 2) set forth to address the extent to which students are spontaneously reminded of distant analogs while generating plausible hypotheses, as well as whether an explicit indication to think of analogous situations can render distant retrievals more likely.

As in the previous experiments, the learning phase was administered by the instructor of an introductory course during a typical lecture, and was presented to participants of both groups as a reading-comprehension activity. Participants received three short stories of between 100 and 140 words and were asked to read them very carefully, since they would have to answer questions about the stories without being able to reread them. While the first and last stories served as distractors, the central story was structurally similar to the target situation to be presented during the subsequent phase. This base analog told the story of a swimming pool whose floor was irreparably fractured, such that the owners planned to fill its 600 $m^3$ with an incompressible material, so as to allow placing a tiled floor over the top face of the pool. The story continued that since no warehouse had a stock of 600 $m^3$ of the same material, the owners ordered 350 $m^3$ of small pebbles and 250 $m^3$ of a comparatively larger type of stones, both of which were being alternatingly thrown to the interior of the pool as the trucks from both companies unloaded their cargo on the sidewalk. The story ended that as the smaller pebbles naturally tended to fill the interstitial space between the stones of the larger type, the purchased material fell short of filling the pool to the top.

The transfer phase was administered by the experimenters during the following class, and presented to participants as a study on hypothesis generation. Before receiving the target situation, participants of the hypothesis-generation condition were handed an instructional material that covered general features of explanatory hypotheses, and illustrated the concept of explanatory hypothesis with two examples, none of them of analogical nature. In contrast, the material handed to the analogical hypothesis-generation condition focused on the use of analogies in explanation, and included examples of both intra and interdomain analogies so as to avoid biasing participants' conceptualization of ideal analogical explanations in any particular direction. Once the time allotted to reading the instructional materials had elapsed, participants of both groups were presented with a target situation pertaining to the domain of chemistry, but which maintained structural similarities with the rather physical situation serving as base analog. The target situation stated that the action of combining 1 L of water with 1 L of alcohol did not yield exactly two liters of solution, but instead 1.9 L (see complete materials in Table 6.2). While participants in the hypothesis-generation condition were simply asked to come up with possible explanations for the presented phenomenon, participants in the analogical hypothesis-generation condition were further asked to think of analogous situations that could suggest potential explanations for the presented situation. Participants of both conditions were encouraged to include as many different explanations as they

**Table 6.2** Materials used in the hypothesis generation and the analogical hypothesis-generation conditions (Trench et al., 2020, Experiment 2)

| |
|---|
| *Base analog*. When the pool of the club got completely fractured, members agreed to fill its volume of 600 $m^3$ with an incompressible material, so as to allow the installation of a tiled floor on its upper face. One warehouse provided 350 $m^3$ of small pebbles, and another store provided 250 $m^3$ of a larger type of stones. Trucks from both warehouses unloaded the stones in the sidewalk, and workers kept filling their wheelbarrows with stones from either pile and throwing them into the pool. To everyone's surprise, the 600 $m^3$ of stones fell short of filling the pool to the top. The reason was that as large stones were mixed with small pebbles, the smaller units tended to occupy the empty spaces left between the larger ones |
| *Target analog*. When two liquid substances are combined, on occasions the volume of the resulting solution does not equal the sum of the initial components. For example, when combining 1 L of alcohol with 1 L of water, the resulting solution does not yield a volume of 2 L, but one of only 1.9 L. What could be the cause of this intriguing phenomenon? |
| *[Analogical] hypothesis-generation task*. Using your imagination, try to provide explanations for why the end volume may not have equaled the sum of the initial volumes of alcohol and water. [Before generating each of your explanations, make an effort to remember other phenomena or situations that are in some sense analogous to the above phenomenon. Do not care much about whether such analogous situations come from Chemistry, or from any other field] |

*Note:* Text between brackets was only included in the analogical hypothesis-generation condition

could think of within the 10 min frame allotted to the explanation task, and were told not to care much about whether the generated hypotheses would likely represent the scientifically accepted account of such phenomenon. Since it is possible that some participants of either group had retrieved the base situation of the first phase but did not transfer the base explanation to the target phenomenon, a post-task questionnaire directly queried participants about whether any of the stories read during the text-comprehension activity had come to mind, even if just briefly, while they were thinking of potential explanations. Participants responding "yes" to this question were further asked to state exactly which story (or stories) had come to mind during the explanation activity. Our dependent measure consisted in whether participants reported having been reminded of the source story during the hypothesis-generation activity. Finally, participants were asked to describe the swimming pool story with as much detail as possible.

Responses to the retrieval questionnaire revealed that only one quarter of participants spontaneously retrieved the base analog in the absence of a prompt to think of analogous cases. More crucially, results revealed that participants who were explicitly prompted to think of analogous cases were more likely to be reminded of the base analog than participants not receiving this prompt, 45.45% vs. 27.69%, $\chi^2$ (1, 131) = 4.45, $p = 0.0349$, $\varphi = 0.18$.

Taken collectively, the results obtained after applying a voluntary search manipulation to a relevant sample of cognitive activities suggest that the effectiveness of a metacognitive disposition to look for relevant analogs depends on the nature of the task at hand: it proved very effective in the case of argumentation, moderately effective in the case of hypothesis generation, and catastrophically ineffective in the case of problem-solving. How to make sense of this differential effectiveness? Far from being confined to the realm of educational interventions, this differential pattern can potentially shed light on the specific way in which general purpose processes as versatile as analogical thinking relate to the real-world activities in which they participate, an aspect receiving virtually no attention within current theorization on analogical retrieval.

## 6.4 Constraining a Model of Voluntary Analogical Retrieval

Even though we tend to agree with the adaptive advantages of continually searching LTM based on the current contents of WM, we have pointed out that the directive character of voluntary search might have the potential to induce the retrieval of analogical sources that would not come to mind in a spontaneous manner. Our results showed that the indication to think of analogous cases proved more effective for some cognitive activities than for others. What dimensions might explain this variation? For a voluntary retrieval prompt to be effective, one precondition that needs to be met is that the cognitive task within which analogical retrieval is embedded does not reliably elicit a conscious search for analogs in and by itself, thus leaving some room for improvement. For those activities which do not naturally invite a deliberate search for analogs, another precondition for a voluntary search manipulation to succeed would be that the mental representations that typically constitute the targets of such cognitive task lend well to a successful memory search. If by their very nature, those representations do not support the extraction of powerful memory cues, the addition of a conscious disposition to carry out a search process might still fail to yield measurable results.

The application of this scheme to the case of argumentation activities can account for the relative success of our voluntary search manipulation. With regard to the first component of this explanatory model, the fact that participants do not generally favor analogical arguments over factual alternatives (Keane & Bohan, 2004) would suggest that an explicit prompt might contribute to initiate a voluntary process that would not be spontaneously triggered in the absence of such prompt. With regard to the second component of this model, it could be argued that the mental representations about which the reasoners have to build a persuasive case tend to be comparatively more "complete" and "well defined" than the open-ended situations that are typically employed as target analogs within the problem-solving literature. In contrast with the representation of an unsolved problem, wherein the relevant aspects can only be clearly distinguished from irrelevant ones in hindsight—that is, after a proper solution has been achieved—the proponent of an argument starts off with a complete representation of the idea she wants to convey. The function of retrieving a suitable analog is not that much to enhance her own representation of the idea, but rather the relatively impoverished representation attributed to the recipient of the analogy. According to O'Keefe and Costello's (2008) computational model of analogical retrieval (see Chap. 3), "connected" representations such as those that are typical targets of argumentation can be more advantageous than relatively less complete representations for the purpose of retrieving superficially dissimilar sources. And besides their completeness, the targets of persuasive argumentation tend to constitute relatively well-learned conceptual structures, on occasions even deserving lexicalized relational categories, which can serve as powerful cues for retrieving further exemplars of such categories (Jamrozik, 2014; Olguin, Tavernini, Pacella, & Minervino, 2017; Raynal, Clement, & Sander, 2018). As an example, the underlying logic of Blanchette and Dunbar's (2000) "zero-deficit" scenario as well as Trench et al.'s (2016) "increasing debt" scenario could arguably be subsumed by the concept snowball, whose metaphoric meaning involves a negative entity which grows in a seemingly geometric fashion, such that at a certain point it becomes impossible to neutralize. Based on these considerations, it makes sense that a meta-cognitive predisposition to base persuasive arguments on analogous cases may have increased the probabilities of retrieving a distant source from LTM.

As opposed to argumentation, for which we advanced the idea that reasoners might not naturally engage in a deliberate search for analogous cases, standard theorization on problem-solving contends that as the operators of open-ended problems are neither explicit in the problem's formulation nor obvious to a non-expert population, the task itself might naturally invite a conscious search for similar situations (Jones & Langley, 2005; Weisberg, 2006), therefore leaving little margin for improvement by means of an explicit prompt to think of analogous situations. To complicate matters more, the targets of problem-solving are, by definition, incomplete; the purpose of analogical retrieval is precisely to transfer knowledge structures that could make the target situation closer to a desired state of affairs. In light of this analysis, it should hardly be surprising that our prompt to look for analogous situations failed to elicit measurable effects.

As stated above, the two-factor model we are adopting posits that the successful application of a voluntary search manipulation to a given cognitive activity requires the joint satisfaction of two preconditions: (a) that the cognitive activity at stake does not naturally trigger a conscious search for analogous situations in a reliably manner, and (b) that the type of representations that typically serve as targets of said activity support the extraction of powerful memory cues for searching LTM. Albeit less dramatic than in the case of argumentation, our prompt to base explanatory hypotheses on analogous situations yielded a significant effect, thus suggesting that even though some participants in the unprompted condition might have spontaneously reacted to the explanation task by consciously seeking for analogous cases, others must have kept trying to obtain a plausible explanation by first principles, without ever resorting to the retrieval heuristic. But what about the second component, namely, the appropriateness of target representations to support the extraction of powerful memory cues out of the target situation? At least with regard to their "completeness," the uncomprehended phenomena that constitute the targets of hypothesis generation could be considered more akin to the targets of problem-solving (i.e., unsolved problems) than to those of argumentation (i.e., well-articulated ideas). If "completeness" were the sole criterion of appropriateness, then participants who were led to search for analogous situations by our explicit instruction would have still failed to extract powerful cues out of the target representation.

How, then, to explain the intermediate effectiveness of our voluntary search instruction in the case of hypothesis generation? We are led to assume that some other dimension of the typical targets of hypothesis generation might compensate for the relative disadvantage of their inherent incompleteness.

One advantageous feature of the type of situations that typically serve as targets of hypothesis generation might relate to their organization into fields of knowledge for which one can foresee neighboring domains that are likely to be subjected to similar laws. For example, the flow of heat between elements of unequal temperature could quite intuitively be related to other physical and/or chemical phenomena such as the transfer of electric charge, radiation, or humidity. In comparison, the typical targets of problem-solving (e.g., how to eliminate a tumor) seem less informative about other fields of knowledge which are likely to host analogically related phenomena. In light of the results of Trench et al.'s (2016) condition in which participants were asked to look for analogous cases within a given set of promising domains, it seems likely that participants' attempts to build plausible hypotheses upon analogous cases could capitalize on their intuitions about which domains of phenomena could be subject to similar laws as that of the target.

On top of this, the structural features of unexplained phenomena seem more likely than those of unsolved problems to be captured by concise linguistic expressions such as relational categories. As posited by several authors (Gentner & Kurtz, 2005; Jamrozik, 2014; Oberholzer, Trench, Kurtz, & Minervino, 2018) schema-governed categories such as *award*, *assault*, or *inoculation* capture the common organizational elements of the exemplars that belong to the same category of events. Upon confronting an instance of "synchronization" whose underlying causes lie beyond our knowledge (e.g., why neighboring clocks tend to oscillate in phase with each other), one can identify situations bearing inferential potential simply by evoking other cases of synchronization such as birds in a flock, menstrual cycles, or pedestrians on a hanging bridge, and assessing whether the explanatory structure of these situations can be productively exported to the target phenomenon at stake. As will be more thoroughly reviewed in the following chapter, providing a relational concept that captures the structure of a target phenomenon can boost the retrieval of particular instances of such concepts (Jamrozik & Gentner, 2020).

To summarize, the present research points to two important aspects overlooked by empirical and computational investigations on analogical retrieval: whether a voluntary search for analogs represents an advantage over spontaneous analogical remindings, and whether this advantage depends on the cognitive activity in which the analogizer is involved. By embedding retrieval into different activities relevant to education, the studies reviewed in the present chapter represent a potential bridge between current algorithms of analogical thinking and the real-world activities whose patterns of analogical retrieval they are meant to reproduce. Further pursuing this line of research might inspire the refinement of current theoretical models, so as to better account for how the mechanisms responsible of analogical retrieval adapt to the rich variety of activities that analogical reasoning can productively subserve.

## 6.5 Conclusions: Potential and Limitations of Voluntary Search

Analogical retrieval can subserve a myriad of cognitive activities, from the most mundane to the most creative and esoteric (Hofstadter & Sander, 2013). Some of these activities do not posit a serious challenge to the cognitive system, either because they do not require interdomain retrieval or because they provide useful cues about how to achieve it. As an example of the former case, when predicting the occurrence of future events, the usefulness of analogical retrieval is circumscribed to the activation of superficially similar analogs, whose inferential power greatly exceeds that of situations maintaining only structural similarity. As an example of the latter case, the activation of the distant domains that constitute the foundation of conceptual metaphors is facilitated by the presence of words that are allude to such domains (e.g., "this relationship is *low on fuel*").

In other educationally relevant activities such as argumentation, abduction, or problem-solving, retrieving distant sources can be determinant. But as reviewed in previous chapters, the structural limitations of our retrieval systems render this type of retrieval very unlikely, especially for those target situations for which we do not have a multitude of distant analogs stored in memory.

The studies reviewed in the present chapter revealed that for two of these educationally relevant activities, distant analogical retrieval can be augmented by a rather simple intervention: explicitly asking participants to think of analogous situations. Given the seeming triviality of this indication, it is surprising that previous studies had not submitted this intervention to experimental manipulation.

Few researchers have maintained a strong interest in the cognitive analysis of both problem-solving and hypothesis generation. Among these, figures as prominent as Herbert Simon, David Klahr, or David Perkins have often conceptualized scientific discovery as a special case of problem-solving, wherein the reasoner searches a space of hypotheses rather than problem states (see, e.g., Klahr & Dunbar, 1988; Perkins, 2000, Simon, 1992). By revealing a dissociation between the effectiveness of our voluntary search manipulation in problem-solving vs. hypothesis generation, the present results might serve to better conceptualize the particular aspects in which these most relevant activities seem to differ.

As with other interventions whose effectiveness was first assessed in the psychological laboratory, follow-up studies should determine whether the present intervention proves successful in real educational environments. The challenges of transitioning from the vacuum chamber of the psychological laboratory to the noisy environment that characterizes real classrooms are proportional to the length and complexity of the intervention at stake. Hence, the relative simplicity of the voluntary search strategy makes it an ideal candidate for being successfully implemented in educational environments. While its application to real-world argumentation tasks could in principle be as effective as its application to hypothesis generation, the fact that direct (i.e., factual) argumentation is often perceived as being as effective as analogical argumentation suggests that voluntary search might hold a comparatively greater promise for enhancing hypothesis generation, at least for scientific or technical topics for which students might not overestimate their ability to come up with a plausible explanation by first principles. At least in the Argentine educational system, an emphasis in the correct scientific account of natural phenomena progressively discourages students from trying to generate their own explanations. The observed effectiveness of voluntary search for hypothesis generation might potentially contribute to enhance students' predisposition to exercise this fundamental epistemic ability.

As opposed to argumentation and hypothesis generation, problem-solving activities proved immune to our explicit prompt to search for analogous situation. We have speculated that this could be due to the multiplicative effect of two factors: that participants may spontaneously attempt this heuristic, making the voluntary prompt somewhat redundant, and that the typical representation of unsolved problems might not be particularly well-suited for extracting powerful memory cues. But if the retrieval of useful sources is the exception rather than the norm, why does the cognitive system persevere in spontaneously resorting to this seemingly maladaptive heuristic, as posited by dominant accounts of problem-solving (e.g., Weisberg, 2006). We speculate that even in settings as unrepresentative of real-life situations as scholarly instruction, the challenges of retrieving a useful problem tend to be less stringent than in psychological experiments. In laboratory experiments, methodological control imposes that there is one and only stored situation that could potentially illuminate the target, and care is taken to prevent participants from episodically linking it to the target. In contrast, in the vast majority of educational scenarios the source problem is often superficially similar to the target and it has been encoded within the same course and the same physical environment, thus constraining the search space considerably. Hence, the fact that students on occasions succeed in accessing source problems within these less stringent conditions might reinforce a natural proclivity to react to new problems by spontaneously searching for analogous situations.

But as stated in several segments of the present monograph, the most interesting cases of creative analogizing are those that are intrinsically more difficult. As opposed to finding idiosyncratic and thematically distant sources during argumentation and hypothesis generation, for which rather "physical" strategies such as iterative search might prove effective, finding interdomain matches for unsolved problems would seem to require that the analogizer goes beyond brute force strategies and attempts to elaborate on the composition of the target representations themselves. Analogous to a shift between a physical (e.g., taking a cold shower) and a chemical (taking antibiotics) approach to a high fever, the interventions to be reviewed in the last chapter of this monograph will assess whether more sophisticated manipulations dealing with the semantics of the target representations can aid reasoners retrieve distant sources during problem-solving, the most refractory of the cognitive activities for which analogical retrieval can play an important role.

## References

Amorín, S., Martínez Frontera, L., Olguín, M. V., & Trench, M. (2017). Spontaneous vs. voluntary analogical retrieval across different cognitive activities. Paper presented at *Analogy2017: The Fourth International Conference on Analogical Reasoning*. Paris, France, July 2017.

Blanchette, I., & Dunbar, K. (2000). How analogies are generated: The roles of structural and superficial similarity. *Memory & Cognition, 28*, 108–124.

Catrambone, R., & Holyoak, K. J. (1989). Overcoming contextual limitations on problem-solving transfer. *Journal of Experimental Psychology: Learning, Memory, and Cognition, 15*, 1147–1156.

Clement, J. (2008). *Creative model construction in scientists and students*. New York: Springer.

Dunbar, K. (1997). How scientists think: Online creativity and conceptual change in science. In T. B. Ward, S. M. Smith, & S. Vaid (Eds.), *Creative thought: An investigation on conceptual structures and processes* (pp. 461–493). Washington, DC: APA Press.

Duncker, K. (1945). On problem solving. *Psychological Monographs, 58*(5), i, Whole No. 270.

Gentner, D., Brem, S., Ferguson, R. W., Wolff, P., Markman, A. B., & Forbus, K. D. (1997). Analogy and creativity in the works of Johannes Kepler. In T. B. Ward, S. M. Smith, & J. Vaid (Eds.), *Creative thought: An investigation of conceptual structures and processes* (pp. 403–459). Washington, DC: American Psychological Association.

Gentner, D., & Forbus, K. D. (1991). MAC/FAC: A model of similarity-based access and mapping. In *Proceedings of the Thirteenth Annual Conference of the Cognitive Science Society* (pp. 504–509).

Gentner, D., & Jeziorski, M. (1989). Historical shifts in the use of analogy in science. In B. Gholson, W. R. Shadish Jr., R. A. Neimeyer, & A. C. Houts (Eds.), *Psychology of science: Contributions to metascience* (pp. 296–325). New York: Cambridge University Press.

Gentner, D., & Kurtz, K. J. (2005). Relational categories. In W. K. Ahn, R. L. Goldstone, B. C. Love, A. B. Markman, & P. W. Wolff (Eds.), *Categorization inside and outside the laboratory: Essays in honor of Douglas L. Medin* (pp. 151–175). Washington, DC: American Psychological Association.

Gick, M. L., & Holyoak, K. J. (1980). Analogical problem solving. *Cognitive Psychology, 12*, 306–355.

Gick, M. L., & Holyoak, K. J. (1983). Schema induction and analogical transfer. *Cognitive Psychology, 15*, 1–38.

Hofstadter, D. R., & Sander, E. (2013). *Surfaces and essences: Analogy as the fuel and fire of thinking*. New York: Basic Books.

Holyoak, K. J., & Koh, K. (1987). Surface and structural similarity in analogical transfer. *Memory & Cognition, 15*, 332–340.

Jamrozik, A. (2014). *The effect of labels on relational retrieval*. Unpublished Ph.D. Thesis, Northwestern University.

Jamrozik, A., & Gentner, D. (2020). Relational labeling unlocks inert knowledge. *Cognition, 196*, 104–146.

Jones, R. M., & Langley, P. (2005). A constrained architecture for learning and problem solving. *Computational Intelligence, 21*(4), 480–502.

Keane, M. T. (1987). On retrieving analogues when solving problems. *Quarterly Journal of Experimental Psychology, 39*, 29–41.

Keane, M. T., & Bohan, A. (2004). Should politicians stop using analogies? Whether analogical arguments are better than their factual equivalents. In D. Gentner, K. Forbus, & T. Regier (Eds.), *Proceedings of the 26th annual conference of the cognitive science society* (pp. 660–665). Austin, TX: Cognitive Science Society.

Klahr, D., & Dunbar, K. (1988). Dual space search during scientific reasoning. *Cognitive Science, 12*, 1–48.

Kurtz, K. J., & Loewenstein, J. (2007). Converging on a new role for analogy in problem solving and retrieval: When two problems are better than one. *Memory & Cognition, 35*, 334–341.

Lombrozo, T. (2006). The structure and function of explanations. *Trends in Cognitive Sciences, 10*, 464–470. https://doi.org/10.1016/j.tics.2006.08.004

Nersessian, N. J. (1992). How do scientists think? Capturing the dynamics of conceptual change in science. In R. N. Giere (Ed.), *Cognitive models of science* (pp. 5–22). Minneapolis, MN: University of Minnesota Press.

O'keefe, D., & Costello, F. (2008). A fast computational model of analogical retrieval (and mapping). In B. C. Love, K. McRae, & V. M. Sloutsky (Eds.), *Proceedings of the 30th annual conference of the cognitive science society* (pp. 2003–2008). Austin, TX: Cognitive Science Society.

Oberholzer, N., Trench, M., Kurtz, K., & Minervino, R. (2018). Analogies without commonalities? Evidence of re-representation via relational category activation. *Frontiers in Psychology, 9*, 2441. https://doi.org/10.3389/fpsyg.2018.02441

Olguin, V., Tavernini, M., Pacella, L., & Minervino, R. (2017). Analogical retrieval mediated by everyday schema-governed categories. Paper presented at Analogy 2017: The Fourth International Conference on Analogy, Paris.

Perkins, D. N. (2000). *Archimedes' bathtub: The art and logic of breakthrough thinking*. New York: Norton.

Raynal, L., Clement, E., & Sander, E. (2018). Structural similarity superiority in a free-recall reminding paradigm. In C. Kalish, M. Rau, J. Zhu, & T. Rogers (Eds.), *Proceedings of the 40th annual conference of the cognitive science society* (pp. 2324–2329). Austin, TX: Cognitive Science Society.

Saner, L., & Schunn, C. D. (1999). Analogies out of the blue: When history seems to retell itself. In M. Hahn & S. Stoness (Eds.), *Proceedings of the 21st annual conference of the cognitive science society* (pp. 619–624). Mahwah, NJ: Erlbaum.

Simon, H. A. (1992). Scientific discovery as problem-solving. *International Studies in the Philosophy of Science, 6*, 3–14.

Spencer, R. M., & Weisberg, R. W. (1986). Context-dependent effects on analogical transfer. *Memory & Cognition, 14*, 442–449.

Sternberg, R. (1990). *Metaphors of mind: Conceptions of the nature of intelligence*. New York: Cambridge University Press.

Trench, M., Oberholzer, N., & Minervino, R. (2009). Dissolving the analogical paradox. Retrieval under a production paradigm is highly constrained by superficial similarity. In B. Kokinov, D. Gentner, & K. Holyoak (Eds.), *New frontiers in analogy research* (pp. 443–452). Sofia: NBU Press.

Trench, M., Olguín, V., & Minervino, R. (2016). Seek, and Ye shall find: Differences between spontaneous and voluntary analogical retrieval. *Quarterly Journal of Experimental Psychology, 69*, 698–712.

Trench, M., Rivas, L. E., Diaz, M. V. & Minervino, R. (2020). Spontaneous and voluntary analogical retrieval during problem-solving and hypothesis generation. In S. Denison, M. Mack, Y. Xu, & B. Armstrong (Eds.), *Proceedings of the 42nd Annual Conference of the Cognitive Science Society* (pp. 1384-1390). Austin, TX: Cognitive Science Society.

Weisberg, R. W. (2006). *Creativity: Understanding innovation in problem solving, science, invention, and the arts*. Hoboken, NJ: John Wiley & Sons.

# Chapter 7
# Boosting Retrieval Via Target Elaborations (the "Late Abstraction Principle")

> Lorenzo was proud of his wife, and wanted his boss Colli to admire her beauty. It wasn't—he thought— the secret pride of the beloved, but that of who possesses a rare and precious object, and who would ultimately wish that others could share his admiration: it was the pride of an *owner*. At some point, he recalled Herodotus' story about the Lydian King Candaules and his minister Gyges. Candaules had arranged for Gyges to watch his wife as she undressed. The Queen got furious, and thus offered Gyges to choose between being executed and murdering her husband. Gyges chose the latter, and went on to marry her. Maybe this comparison, Lorenzo thought, would extend towards an analogous final disaster: just as Gyges ended-up becoming the Queen's husband, so his boss could become the lover of his wife.
>
> Alberto Moravia, in *The Leopard Woman*

As reviewed in Chap. 5, promoting a more abstract encoding of the base analogs renders such sources more likely to be retrieved during a later encounter with a superficially dissimilar target. The limitation of this approach is mainly one of applicability, since it provides no solution to the retrieval of situations learned in conditions that were not especially engineered to enforce this kind of elaborations. While source abstraction could readily be used to teach educational contents in ways that would promote its later utilization across conceptual and contextual barriers, the vast majority the educational and extra-educational contents that we have acquired during our lifetime are not likely to have benefitted from this type of encoding, and therefore might fall prey to the competence limitations discussed during the first part of the present monograph. Chapter 6 reviewed a recent set of studies aiming to improve access to suboptimally encoded situations by way of giving participants a brief instruction about what constitutes a good analogy in general, and asking them to reflect on a target situation by deliberately trying to evoke analogous situations from memory. The effectiveness of this rather "physical," brute-force approach to augmenting retrieval proved very successful during argumentation, moderately effective during hypothesis generation, and ineffective during problem-solving, probably due to the fact that the mental representations of unsolved



M. Trench, R. A. Minervino, *Distant Connections: The Memory Basis of Creative Analogy*, SpringerBriefs in Psychology,
https://doi.org/10.1007/978-3-030-52545-3_7

problems neither lend well to the extraction of powerful memory cues nor do they point to promising knowledge domains wherein to circumscribe the memory search. How, then, to find a subtler solution to the retrieval of suboptimally encoded sources?

The cited excerpt from Alberto Moravia's posthumous novel "The Leopard Woman" presents a hint about such possibility. Note that Lorenzo—the main character of the novel—was proud of his wife and wished that Colli, his boss, could admire her beauty. But before the analogous story about King Candaules was brought to his consciousness from the bottom of his memory, he had just articulated a more abstract reformulation of his current situation: "It wasn't the secret pride of the beloved, but that of who possesses a rare and precious object, and who would ultimately wish that others could share his admiration: it was the pride of an *owner*."

Was Moravia intuitively grasping the provocative idea that it was the Lorenzo's reformulation what facilitated access to the Candaules episode? Until recently, the main mechanism posited to account for the retrieval advantage of deriving abstract schemas was the *learning to encode* principle, according to which if a base analog is indexed as a case of a more abstract formulation, this schema could be available and become activated during the processing of the target, thus serving as a powerful cue to retrieve base analogs that had been indexed as cases of this general formulation (Medin & Ross, 1989). A funny anecdote during Douglas Hofstadter's trip to the Grand Canyon's North Rim (Hofstadter & Sander, 2013) illustrates this possibility. While he and his wife were being captivated by the huge chasm, their fifteen months old son Danny was riveted by a few ants and a leaf on the sandy ground, fifty feet from the canyon's edge. The irony of this situation could plausibly have been captured by Hofstadter in terms of "a trivial side-show more fascinating than the main event," thus paving the way for being reminded of the ants' episode upon an encounter with a similar situation wherein a four-year-old girl had just received a lovely Christmas present, but she paid no attention to it and instead concentrated on the sparkling wrapping paper. But what would have happened if, as it was probably the case in Moravia's story, the source episodes had not elicited an abstract formulation at the time they were encoded in LTM? Could target abstraction conceivably operate backwards in time? On the "learning to encode" account, target abstraction could only be effective in cases wherein the sources had already been encoded in abstract terms.

As reviewed in Chap. 5, there was another theoretical possibility by which more abstract sources could be retrieved more easily than non-abstracted ones during the processing of a superficially dissimilar analog. In those cases wherein the base and the target analog differ in surface features but maintain an identical system of relations, any intervention aimed at removing surface features from the base analog or lessen their importance might facilitate its retrieval when confronted with a distant target, since surface mismatches will no longer obscure a good structural match. As shown in MAC/FAC simulations of forward transfer (see Sect. 5.3), the advantage of abstracted targets emerges during the MAC stage as a natural consequence of vector normalization: Given that a fixed amount of weight is distributed among all the non-zero slots of the content vectors, the removal of object attributes entails allocating more weight to relational predicates, thus increasing the dot product between the now abstracted source and an analogous target.

Gentner and colleagues (Gentner, Loewenstein, & Thompson, 2004; Kurtz & Loewenstein, 2007) reasoned that as dot products are commutative, then the strength of a source-target match such as those computed by the MAC stage should also increase if the abstraction process were applied to the target, instead of the base. Furthermore, the removal of object attributes from the target would diminish the retrieval of stored items sharing superficial but not structural features with the target. They called this insight "the late analogical abstraction principle," and set forth to gauge experimental evidence about its psychological reality.

## 7.1 Asking Participants to Compare Two Analogous Targets

In a series of studies, Gentner, Loewenstein, Thompson, and Forbus (2009) had negotiators of different degrees of expertise either compare or separately read two instances of "contingent contracts," an optimal negotiation strategy for cases in which the parties have diverging estimations of a critical future event, and thus leave open the terms of the contract as a function of the outcome of said event. For example, a theater owner and a producer are at an impasse over negotiating costs associated with mounting a new production: The producer anticipates sold-out shows and demands a correspondingly high price, whereas the theater owner has more modest expectations and, thereby, wishes to minimize expenses. Instead of compromising on a midpoint between their initial proposals, they can agree that the producer will get 60% of the revenues if more than 70,000 seats are sold, but only 45% of the revenues if selling fewer seats. In a first experiment involving negotiators with an average of 15 years of experience, the instructional material of the comparison condition presented two analogous contingent agreements with mismatching surface and contextual features, followed by the question: "What is going on in these negotiations? Think about the similarities between these two cases; what are the key parallels between the two negotiations? Please describe the solution and say how successful you think it is." In contrast, the separate cases condition received the two cases separately, and had to answer the following question after each of the cases: "What is going on in this negotiation? Please describe the solution and say how successful you think it is." Upon answering these questions, participants of both conditions were asked to recall an example like those they have just read, as well as to state the source of their example: "own experience, colleagues/partners, popular press, prior class, other." Finally, each participant was paired with another participant of the same condition in a face-to-face negotiation, and experimenters assessed whether the contingent contract strategy was employed. Participants in the comparison condition scored higher than participants in the separate-cases condition on all three dependent measures: case descriptions (a measure of principle understanding), quality of contract (a measure of transfer), and quality of retrievals from LTM. Subsequent experiments generalized the above findings to a much less expert population (students in a Masters of accounting program, Experiment 2) and to negotiation strategies different from contingent contracts (trade-offs, Experiment 3). While 54% of the reported cases were business situations, the remainder were examples from daily life.

As compared to the production paradigm implemented by Blanchette and Dunbar (see Chaps. 3 and 4), the fact that the experimenters asked participants to state the source of their examples rendered them comparatively more verifiable as genuine retrievals. However, as recognized by the authors, there still remained the possibility that some of the sources reported by negotiators were partially or wholly constructed at recall time to match the probes.

To control for the effect of analogy fabrication, Gentner et al. (2009, Experiment 4) resorted to a two-phase cued recall paradigm akin to the one employed in the story reminding tradition (see Chap. 2). During the second phase of the procedure, participants of the comparison and the separate-cases conditions were presented with the same two structurally similar cases of contingent contracts and the same sets of questions as in the homonymous conditions of Experiments 1–3. Upon answering these questions, participants in the comparison condition were asked to recall an example presented earlier in the experimental session that best matched the two cases they had just read on the previous page. Participants in the separate cases condition were asked this question right after reading each of the presented targets and answering its comprehension questions. During the first phase of the procedure, participants had received the critical source analog embedded within six other stories: one surface match for each of the targets, and four unrelated fillers. Albeit far from ceiling performance, participants in the comparison condition were more likely to report the source analog than in the separate cases condition (27% vs. 6%), and less likely to report the surface match (30% vs. 78%). Hence, both in naturalistic conditions and when given a controlled set of source analogs, comparing two analogous targets increased the likelihood of a relational retrieval. However, the combination of a cued recall instruction and a rather narrow set of candidate sources might still have allowed some participants to compare the targets to the sources in a serial fashion, thus circumventing the computational challenge of more realistic conditions wherein the source analog is buried in LTM among a myriad of potential competitors. Furthermore, would the retrieval benefits of target-comparison stretch to problem-solving activities, i.e., the particular task that has proven immune to the more direct and austere intervention of asking participants to think of analogous situations?

Kurtz and Loewenstein (2007) had taken aim at this particular challenge. During a first phase of the procedure, participants of two groups were asked to read Gick and Holyoak's (1980) military story very carefully, in order to be able to retell it in their own words. After handing in these study materials, participants in the comparison condition were provided with two unsolved problems analogous to the military story, but superficially dissimilar from it as well as from each other: the radiation and the Red Adair problems. In this last problem, a fire in an oil well was so severe that it could only be suffocated by placing a large quantity of fire retardant at the base of the well, with the problem being that there was no single hose large enough to carry the quantity of foam needed to extinguish the fire.

After completing a matching task between five corresponding elements of both problems, they were asked to propose solutions to both problems under the consideration that they admitted a common solution. The employment of the convergence strategy under this condition (54%) was higher than under a baseline condition receiving only the tumor problem (15%). By adding a separate targets condition in which both targets were sequentially confronted, Experiment 2 demonstrated that the transfer advantage of the target-comparison strategy in the previous experiment did not rest on the fact that participants were exposed to two targets instead of one. In turn, Experiment 3 added a further control wherein participants received and compared the two target situations but without having received the military problem during a previous phase. The fact that the target comparison group outperformed this latter control condition confirmed that the observed advantage of the target-comparison strategy stemmed from retrieving the source analog from memory, rather than from enabling participants to solve the problems by first principles.

If the cued recall paradigm used by Gentner et al. (2009, Experiment 4) was suspicious of trivializing the challenge of finding a needle in a haystack, so it could be considered the fact that in Kurtz and Loewenstein's procedure the retrieval task was presented immediately after the processing of a source analog that was not even embedded between distracter stories. Conceived as a natural extension of Kurtz and Loewenstein's study, the last experiment of Gentner et al. (2009) set forth to introduce a stronger contextual separation between the learning and the transfer phases. Albeit displaying lower retrieval rates overall, participants who compared two analogous targets were more likely to propose convergent solutions (31%) than participants who read the problems separately (5%).

Taken collectively, the results from the above thread of related studies attest to the psychological reality of the late abstraction principle, that is, the thesis that by arriving to a more abstract encoding of a target situation, the reasoner can increase her likelihood of retrieving distant analogs that would otherwise remain inert in LTM.

As opposed to the "learning to encode" principle that inspired the source abstraction interventions reviewed in Chap. 5, the late abstraction principle opens a promising avenue for retrieving base situations learned within contexts that were not specially engineered to highlight their underlying structures, and which may represent the vast majority of the learning conditions that people experience within and outside instructional settings (Loewenstein, 2010). But the problem resides in that the potential implications of the late abstraction principle contrast sharply with the applicability limitations of its concrete behavioral implementation: the target-comparison strategy. As some readers might have already realized, the target-comparison strategy inevitably requires that learners are externally provided with a further analogous situation for any target situation they are to confront. In other words, they will need to carry Jeff Loewenstein in their backpack!! One of the pressing questions in our own research program is thus: can learners capitalize on the late abstraction principle, but in truly autonomous ways? The remainder of the chapter reviews successful and failed attempts to generate portable interventions based on the late abstraction principle, and finally discusses our intuition—shared with Loewenstein and colleagues, but by no means undisputed—that late abstraction affords the retrieval of suboptimally encoded learning.

**Fig. 7.1** A humorous representation of the applicability limitations inherent in Kurtz and Loewenstein's (2007) target-comparison strategy

## 7.2 Comparing the Target to a Disanalogous Problem

Conceived as a minimal variation to the target-comparison strategy, Minervino et al. (2017), Experiment 1 indication to compare the target to a random disanalogous problem represents our first attempt to devise a portable intervention based on the late abstraction principle. We conjectured that the activity of comparing disanalogous problems might still lead participants to extract abstract cues that could eventually subserve the retrieval of distant analogs. Our hypothesis was that an intuitive analysis of both problems in terms of their structural components (i.e., their goals restrictions, operators, etc.) would frame the establishment of correspondences between the concepts that play such roles in each of the problems. This could in turn result in conceptualizing commonalities and differences between those fillers, eventually giving rise to appropriate cues for retrieving distant analogs. As an example, consider that before attempting to solve the radiation problem, participants had to compare such problem to Duncker's (1945) *Candle* problem, where participants are presented with candles and a matchbox-size box filled with tacks, with the challenge of attaching the candle to a wall in such a way that the wax of the candle would not drip onto the floor. When comparing these two problems, an analysis of their respective restrictions could lead to conceptualize them as two instances of "avoiding to affect nearby elements." On the other hand, an analysis of the permitted operators could lead to conceive them as instances of "using a force vs. using office items to reach an objective." Similarities and differences of this kind tend to be more abstract than the particular elements in which they originate. Hence, including them in the representation of a target problem could potentially increase the retrieval of distant sources by means of providing additional and more appropriate cues. In case it did, this alternative route to abstraction would represent a more portable strategy than comparing the target to an externally provided analog, since students could randomly choose any of a potentially indefinite number of problems to be compared against the target problem be solved.

To assess the usefulness of this intervention, we compared the proportion of convergence solutions given to the tumor problem by participants who were asked to compare the radiation to a disanalogous problem against two control conditions: one that was identical except for the fact that participants were simply asked to solve the tumor problem, and a second one wherein participants were asked to compare the radiation against a disanalogous problem, but without having previously read the military problem during a contextually separated phase of the procedure. The disanalogous problem used for comparison was an adapted version of the *Weigh the Asteroid* problem used by Chen, Mo, and Honomichl (2004), for a description of this study, see Chap. 4. In the present version, four scientists arrive by boat to a remote island where a mysterious explosion has been recorded. They find a large asteroid and plan to load it on their boat, but one of the scientists reads from the safety warnings that the boat can safely carry up to 800 kg. They have a scale in the boat, but it can weigh objects of at most 200 kg, which is insufficient to weigh the asteroid. The problem asks participants to suggest possible ways of using the extant scale to determine the weight of the asteroid, but without breaking it into smaller fragments.

An analysis of the comparisons produced in the comparison condition showed that participants who managed to construct abstract schemas had greater probability of producing a convergence solution to the tumor problem than those who failed to produce such schemas. However, the fact that less than 25% of participants could generate successful schemas prevented the comparison group from outperforming participants in the control conditions.

## 7.3 Inventing an Analogous Target

In a further attempt to devise portable ways of inducing a more abstract encoding of the target analog (Minervino et al., 2017, Experiment 2) we turned our attention to the invention of analogous problems. The construction of isomorphic problems had been considered a stringent test of learners' comprehension of different types of mathematics and algebra problems, as well as of their ability to relate them to the domains of real-world situations to which they can apply (see, e.g., Nathan, Kintsch,

& Young, 1992). In a study using probability problems, Bernardo (2001) found that the activity of constructing an analogous problem can be beneficial for later transfer, even surpassing the problem-comparison task in efficacy. The advantage of generating analogous exemplars reaches beyond the domain of mathematics, and even of problem-solving in general. For example, in an unpublished study reported by Dunbar (2001), participants who read fable-like stories with the task of generating analogous episodes were more likely to retrieve them during the processing of analogous stories lacking surface similarities.

Based on the above studies, we ventured that the activity of generating an analogous situation would direct participants' attention towards the structural features of the target problem being processed, thus enhancing the retrieval of potentially useful source analogs (i.e., the late abstraction principle). In keeping with the radiation problem as an example of a possible target analog, a person could confront the challenge of fabricating an analogous problem by tentatively replacing the patient's stomach by the bone of a fruit, guided by a surface resemblance such as their spatial location inside an object. Next, she would have to relate the bone to another entity (e.g., a mold inside the bone), whose *functionally relevant attributes* (Keane, 1985) would allow it to play in the fruit scenario the same role played by the tumor within the radiation problem (threatening the rest of the fruit). The selected entity, in turn, constrains the selection of possible operators to those that would eliminate the focus of the problem when applied at the required intensity (e.g., by boiling the whole fruit), but at the cost of harming a critical element that needs to be preserved (e.g., the texture of the pulp).

As illustrated by the above example, the derivation of an analogous target problem in a non-formal domain seems to demand a deep and systematic exploration of the relational structure of the problem. In terms of portability, the advantage of target construction over target comparison is easy to foresee: participants would be able to apply it to any potential target situation without the need to be externally provided with a second target analog.

To assess the effectiveness of this instructional intervention, we compared the proportion of convergence solutions given to the tumor problem by participants who were asked to construct an analogous problem against two control conditions: one that was identical except for the fact that participants were simply asked to solve the tumor problem, and a second one wherein participants were asked to construct an problem analogous to the tumor, but without having previously read the military problem during a contextually separated phase of the procedure. In line with our expectations, participants in the fabrication condition outperformed participants of the standard transfer condition in producing convergence solutions to the radiation problem. Given that participants who had to invent an analogous target without having previously received a base analog did not outperform participants in the standard transfer condition (10% vs. 10%), the increased performance in the invention condition could not be attributed to an effect of target invention on the ability to solve the target problem from first principles. Thus, the activity of constructing a novel unsolved target problem at test time seems to improve transfer by fostering the retrieval of the base story and its convergence solution.

In order to have an estimate of how our target invention strategy fares against target comparison in terms of transfer efficacy, we compared our results with a prior replication of Kurtz and Loewenstein (2007) that we had carried out with the same population of participants. Even though target invention is potentially more portable, it was somewhat less effective than target comparison (25.71% vs. 34.29%). A cross-measures analysis revealed that participants who managed to generate an analogous problem were highly likely to transfer the convergence solution to the radiation problem. Therefore, the slightly lower transfer rate of target construction as compared to target comparison likely originates in the fact that participants found it quite difficult to fabricate an analogous target.

To understand the sources of this difficulty, it should be noted that most prior interventions involving problem construction had been developed as ways of promoting and/or testing the acquisition of concepts and procedures in mathematical subdomains such as algebra (Nathan et al., 1992), arithmetic (Lampert, 1986; Rudnitsky, Etheredge, Freeman, & Gilbert, 1995), or probability (Bernardo, 2001). Generating analogous problems within formal domains seems to allow a great freedom for choosing the objects that instantiate the given quantities of the problem (e.g., apples, baskets, etc.), as well as by the availability of easily accessible actions that can potentially reinstantiate the mathematical operations represented by the base relations (e.g., replacing *putting* by another instance of adding, such as *buying*, and replacing *removing* by another instance of subtracting, such as *selling*). In contrast, generating analogous problems outside the realm of formal disciplines might require the deployment of more aggressive re-representation mechanisms, as well as a more creative exploration of our sometimes impoverished knowledge about certain domains (Rozenblit & Keil, 2002). To illustrate, the generation of a problem like *Red Adair* on the basis of the radiation problem could have begun by shifting the goal of the problem from destroying a tumor to suffocating a fire. Once this new goal was set, the remaining elements of the problem have to be adjusted in a coordinated fashion, a process of multiple constraint satisfaction that can on occasions be rather taxing (Hofstadter, 1985; Ward, Smith, & Finke, 1999). While the replacement of rays by fire retardant is rather straightforward, it is less easy to envision target elements that could fill the role played by the surrounding tissues in the source domain. If the analogous problems invented by experienced scholars for research purposes sometimes suffer from structural mismatches, it is not surprising that participants often failed to carry out these replacements in a complete and coordinated manner. Despite these intrinsic difficulties, it is conceivable that training participants in the generation of analogous problems could render this strategy somewhat easier to implement. In view of the greater portability of target construction as compared to the ad hoc provision of a second analogous target, it seems to us that any investment in enhancing this cognitive ability would promise to pay off well.

## 7.4 Constructing Abstract Conceptual Representations of the Target

As reviewed in Chap. 5, Mandler and Orlich (1993) had shown that the provision of personalized feedback from the experimenter in the context of an explicit training can help participants generate schemas at different levels of abstraction, ranging from concrete descriptions to abstract schemas to vacuous statements. The application of these techniques to the military story rendered this episode more retrievable during later exposure to the radiation problem. In a recent study (D'Angelo & Trench, 2020, Experiment 1) we set forth to assess whether attempting to distill an abstract representation of unsolved problems at retrieval time would, by virtue of the late abstraction principle, help participants retrieve distant analogs from memory. On top of the challenge of aiding participants in generating appropriate schemas for unsolved problems—recall that the solution to a problem typically sheds light on the identification of central vs. irrelevant information—we were interested in employing a self-administered type of training, for the sake of both applicability and replicability. Upon confirming that participants of our student population seldom arrived at abstract schemas of the type described by Mandler and Orlich, we presented participants of our abstraction condition with an instructional material focused on arriving at abstract representations of situations or problems by means of replacing concrete nouns such as *message* or *policeman* by indefinite pronouns such as *something* or *somebody*, respectively. In line with the idea that removing object properties from the target representation could enhance distant retrieval by suppressing surface content in the favor of relational structures, we predicted that the deployment of indefinite pronouns would render distant retrieval more likely. However, results showed that even though participants somehow managed to apply this technique to redescribe the radiation problem, the retrieval of the source situation did not prove statistically superior to that of a control group who was not asked to elaborate on the problem's representation. We speculate that the replacement of nouns by indefinite pronouns might have suppressed, not only irrelevant features of the problem's entities, but also properties of such entities that are relevant for the role they play in the described situation (Falkenhainer, 1990). On the other hand, the replacement activity might have been so mechanical that students could have performed it in a localist manner, that is, without ever bothering to maintain the whole structure of the target active in working memory. If this were the case, the sequential simultaneous activation of small arrays of propositions having no more integration than their co-occurrence in a sentence might have been ill-suited for supporting the retrieval of complex analogous structures from LTM. A handful of computer models of analogical reasoning had stressed that the activation of large propositional structures in working memory is neither psychologically plausible (Hummel & Holyoak, 1997) nor desirable for interdomain retrieval (see, e.g., Finlayson & Winston, 2006). In view of the above considerations, a more efficient search for superficially dissimilar analogs as complex as those typically used in the problem-solving literature would consist in fostering a more concise representation of the target problem.

In a subsequent experiment, D'Angelo and Trench (2020, Experiment 2) assessed the effectiveness of asking participants to briefly characterize the role played by key objects within the particular situation depicted by the target problem. To this end, two groups of participants read an adapted version of the Fire Chief problem plus two distracter stories in the context of a study on reading comprehension. After a brief delay and a contextual separation, both groups were invited to participate in a study on problem-solving. In the abstraction condition, participants received a brief instructional material on how to describe central objects of a problem according to the situation-specific role they played in relation to the other elements of the problem. Next, they were given the radiation problem with the instruction to generate relational descriptions of the tumor and the rays. Before proposing potential solutions, they were asked to think of other known situations involving objects amenable to similar descriptions. Participants in a control condition received the tumor problem after completing an unrelated task whose time demands were similar to those of the instructional material of the abstraction condition. As opposed to the results of our indefinite pronouns manipulation, the retrieval rates of the abstraction condition (43.6%) doubled those of the control condition (20.7%).

More in line with Mandler and Orlich's (1993) original idea of generating abstractions from a given situation, we are currently developing a self-administered training schedule aimed at helping participants generate fully articulated abstract schemas at the maximum possible degrees of generality that could be reached without sacrificing essential aspects of a problematic situation.

## 7.5 Constructing Idealized Visual Representations of the Target

As an alternative to focusing on the conceptual abstraction of a target situation, in a recent study (Trench, Tavernini, & Goldstone, 2017) we turned our attention to the *perceptual* idealization of the target analog as a portable strategy potentially capable of enhancing the retrieval of distant analogs at retrieval time. As reviewed in Chap. 5, Goldstone and colleagues had demonstrated that even though concrete instantiations of an abstract principle are more beneficial than idealized representations for inducing an initial understanding of such principle, the reverse is true when it comes to recognizing a semantically different analogous situation as a further instance of such principle.

In order to assess whether a comparable transfer advantage can be obtained by inducing a more idealized representation of the target analog at retrieval time, Trench, Tavernini, and Goldstone (Experiment 1) had three groups of participants learn how to use the $r_1.t + r_2.t = d$ formula for solving a "collision" problem in which a plane and a helicopter traveled towards each other at different speeds. After using instruction on another algebra topic (combinatorics) as a means for contextual separation, participants were presented with a "work" problem in which they had to calculate the time that two painters would need to jointly paint a wall, given the times that each of them would have needed to paint it on his own. Before being asked to actually solve the problem, the two simulation groups were presented with a set of manipulatives and were asked to carry out an approximate representation of the situation described by the target problem as it unfolded from the outset of the painting until the moment when the wall got completed. Participants in the *concrete* condition received a photograph of a horizontally laden wall and two smaller 1.6 in. × 0.75 in. rectangles printed with drawings of painters (see Fig. 7.2). In order to get a record of the specifics of each participant's simulation, the next page included three similar walls meant to represent three different snapshots of the dynamic simulation they had just performed. Upon receiving four additional paper painters (two gray and two black) and a glue stick, they were allotted 2 min to produce a record of the mental simulation they had just performed by means of sticking two painters onto each wall in a manner faithful to the locations of each of the painters at three different moments: (1) at the exact moment when they started painting [top wall], (2) at an intermediate stage of the process [center wall], and (3) at the exact moment when the painting job got was completed [bottom wall]. Once the time allotted to this activity had elapsed, participants were given 5 min to solve the problem by any means. The procedure followed by the idealized simulation group was identical to that of the concrete simulation condition, with the difference that the manipulatives used during the simulation were relatively more abstract. While the wall consisted of a white 6.37 in. × 1.84 in. sized rectangle, the two painters

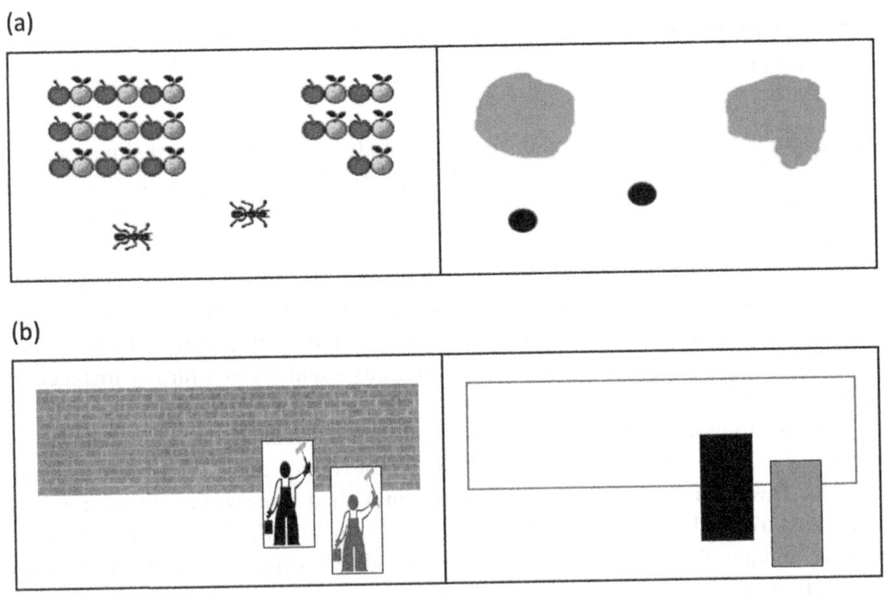


**Fig. 7.2** Manipulatives employed for the carrying out concrete (**b**) or idealized (**a**) representations of the target problem (From Trench et al., 2017)

were represented by 1.6 in × 0.75 in. sized gray/black paper rectangles. The no simulation condition received neither the manipulatives nor any indication to carry out a dynamic simulation of the situation models of the problem.

The spontaneous transfer rate in the idealized condition (79%) was reliably greater than those obtained in the concrete (39%) and in the no simulation conditions (51%). These results suggest that idealized representations were more advantageous than concrete representations for eliciting the retrieval and transfer of the base problem's solution. More crucially, the fact that the idealized simulation condition also outperformed the no simulation condition suggests that that the observed advantage of idealized simulations is not simply an artifact of the potentially detrimental effect of concrete simulations. Hence, it seems to represent a viable instructional intervention.

An alternative explanation, however, could posit that the concrete representations of the painters might have invited a dynamical representation that was inconsistent with the "convergent" representation that characterized the base problems. If the concrete simulation of the painters' activity recruited a "socially laden" representation in which the painters advance in parallel fashion—e.g., to talk to each other—rather than in the more transfer-appropriate "converging" motion, this idiosyncratic accidental feature could have contributed to their inferior transfer performance. In order to assess this possibility, we sorted participants' representations as "convergent" vs. "non-convergent" according to the way in which they had glued the painters onto the three walls that were meant to record three snapshots of how participants intuitively imaged the process as it unfolded over time. This analysis revealed a nonsignificant trend towards a greater use of the convergent representation in the concrete simulation condition, thus discarding a harmful effect of the socially laden interpretation.

Another alternative explanation for the superiority of target representations for eliciting correct solutions to the work problem could be that such advantage was originated, not in the benefits of our idealized materials for analogical transfer (as posited here), but rather in their potential to promote a better understanding of the target problem in its own terms, thus leading to a higher probability of solving such problem by first principles. According to various authors (see Belenky & Schalk, 2014 for a discussion) learning is facilitated when representations convey the minimum detail that is necessary to grasp the quantitative structure of a problem. As an example, the removal of potentially distracting irrelevant features like the quasi-regular pattern of the bricks or the left vs. right handedness of the painters could have helped participants build a more accurate representation of the temporal dynamics of the problem (e.g., the different speeds of each painter), which may serve as a secure foundation from which to control the accuracy and soundness of algebraic manipulations (Minervino, Trench, & Oberholzer, 2009). In order to assess how the concrete and idealized simulations enforced in Experiment 1 impacted the raw probabilities of solving the target problem in a non-analogical fashion, in a subsequent experiment three groups received the same materials and instructions as in the concrete, idealized and no simulation conditions of the previous experiment, but without having been exposed to the base problem during a prior phase. Given that the simulation manipulation did not elicit an effect in the absence of a base analog and its solution, the advantage of the idealization condition obtained in the analogical conditions of the previous experiment stems from an increased retrieval of the distant base analog, and not from a beneficial effect of idealized simulations for solving the target problem by first principles.

Taken collectively, these results suggest a parallelism between the abstraction that takes place during tasks like problem comparison or problem construction and the kind of perceptual idealization induced by our manipulation. Akin to the advantage of abstract retrieval cues in activities like problem comparison or problem construction, we conjecture that idealized representations of the target problem are (1) perceptually more similar than their alternative concrete representations to any base representation, and (2) less likely to evoke spurious remindings that could eventually outcompete the critical base analog. The present results thus contribute to enlarge the empirical basis of the late abstraction principle, while at the same time broadening its scope so as to include a perceptual dimension that has not been thus far discussed in the literature.

## 7.6 Computer Simulations of the Late Abstraction Principle

Gentner et al. (2009) reasoned that as retrieval depends on the degree of match between the stored items and the memory probe, the beneficial effect of schema abstraction that was consistently obtained during source encoding should also apply when elaborating on the target analog at retrieval time. Assuming MAC/FAC to be an adequate model of how people retrieve analogous cases, the authors hypothesized that a more abstract representation of the target analog could potentially augment distant retrievals in two different ways. On the one hand, given that the total weight of the vector representation of a given situation is distributed among the units of its content vector (i.e., unit-vector normalization), removing surface features from the target will entail allocating more weight to its relational structure. Therefore, the dot product between a distant analog and the schema extracted from the target being processed will be higher than that between such base analog and the original, non-abstracted target. On the other hand, as more abstract target representations have fewer specific object attributes and other low-level features, they will attract fewer mere-appearance matches that could compete with thematically distant analogical matches.

To obtain a demonstration of this idea, Gentner et al. (2009) had MAC/FAC (see Chap. 2) run on a subset of seven story sets from Gentner, Rattermann, and Forbus (1993). As in the simulations of the forward transfer effect (see Chap. 5), abstract schemas were obtained via running SME on two analogous stories having the same relational structure but different entities and contextual features. The MLP of MAC/FAC was populated with the analogical match, the mere-appearance match and the first-order relational match from each of the seven sets, plus unrelated stories. During the MAC stage, the use of non-abstracted stories as probes led to the selection of a majority of mere-appearance matches, with the analogical match being rarely retrieved. Having no better match to choose from, the FAC stage returned the mere-appearance matches in nearly 100% of the cases and analogical matches in less than 15% of the cases, a pattern similar to that of previous behavioral studies and MAC/FAC simulations. When the schemas were used as probes, the MAC stage typically retrieved both the analogous story and the mere-appearance match of the corresponding set, thus setting the stage for FAC to select the analogous match in 86% of the runs, and the mere-appearance match in 29% of the cases (see Table 7.1).

As in the MAC/FAC simulations of the forward transfer effect, a second simulation added literally similar items to the memory set in order to assess whether the obtained pattern would change. For the concrete target probe, MAC retrieved both the literal and the mere-appearance matches in nearly all runs, and no other matches. FAC then rejected most of the mere-appearance matches in favor of the literal matches, leading to the retrieval of the literal similes in 100% of the runs and of mere-appearance matches in 29% of the runs. When the schemas were used as targets, the MAC stage retrieved 100% of the literally similar matches, 86% of the analogy matches, and 71% of the mere-appearance matches. The FAC stage then rejected most of the mere-appearance matches, thus retrieving the literally similar matches in 71% of the cases, the analogy matches in 86% of the cases, and the mere-appearance matches in only 14% of the cases.

Results from MAC/FAC simulations show that when non-abstracted stories are used as probes, the algorithm preferentially retrieves literally similar items when available, and mere-appearance matches when no literally similar items are stored in memory. The use of abstract cues increases the proportion of purely analogous retrievals, but without threatening the retrieval of literally similar analogs when they are available. Taken collectively, the MAC/FAC simulations carried out by Gentner et al. (2009) showed how the allotment of more weight to relational predicates can diminish surface competition, thus accounting for the retrieval advantage of target abstraction observed in behavioral studies.

**Table 7.1** Retrieval of different types of similarity matches using concrete-level situations versus abstract schemas as cues (From Gentner et al., 2009)

| | LS match | | Analog | | MA match | | FOR match | |
|---|---|---|---|---|---|---|---|---|
| | MAC | FAC | MAC | FAC | MAC | FAC | MAC | FAC |
| First Sim. | | | | | | | | |
| Exemplar | | | 0.14 | 0.14 | 1 | 0.86 | 0 | 0 |
| Schema | | | 1 | 0.86 | 0.86 | 0.29 | 0.29 | 0.14 |
| Second Sim. | | | | | | | | |
| Exemplar | 1 | 1 | 0 | 0 | 1 | 0.29 | 0 | 0 |
| Schema | 1 | 0.71 | 0.86 | 0.86 | 0.71 | 0.14 | 0.14 | 0 |

*LS* literal similarity, *MA* mere appearance, *FOR* first-order relations, *Sim.* simulation

## 7.7 Can Late Abstraction Aid in Retrieving Suboptimally Encoded Sources?

Most of the excitement about interventions taking place at recall time stems from the assumption that they can aid participants in retrieving suboptimally encoded knowledge from LTM (see Gentner et al., 2009; Minervino et al., 2017; Trench et al., 2017). Even though neither our procedures nor those of Gentner and colleagues encouraged an abstract encoding of the sources, results are silent about the extent to which the sources retrieved by participants in the target abstraction conditions had received an abstract processing during their initial encoding. It could be argued that, as participants' spontaneous encodings of the base analogs vary along a continuum from the very shallow to the very deep, the late abstraction manipulations were only able to aid retrieval in those cases where the base analog happened to receive an abstract encoding, akin to the one generated later during the processing of the target.

Needless to say, empirically demonstrating the retrieval of suboptimally encoded sources would require assessing the nature of an individual person's particular encoding without altering it in any way, something that might be difficult to achieve. In the meantime, such discussion can benefit from gaining further precision with regard to the different ways in which base encodings could be suboptimal, the mechanisms by which such suboptimally stored knowledge could eventually be retrieved, and the possible role of late abstraction in facilitating these particular kinds of analogical retrieval.

One slightly suboptimal way of encoding the base analogs could include its unabstracted relational structure *plus* the specific superficial features, as opposed to the combination of relational structure plus little or no superficial aspects. To the extent that the architecture of MAC/FAC resembles our memory systems in meaningful ways, then the removal of object predicates from target representations should suffice to augment the retrieval likelihood of the above type of "suboptimally" encoded sources, either via diminishing surface competition and/or increasing the relative weight of relations. Target interventions like problem comparison or problem construction could potentially aid the retrieval of this kind of base analogs. A precondition for this to occur, it should be noted, is that the systems of relations of the source and the target are initially identical. Gentner et al.'s (2009) MAC/FAC simulations of backward transfer represent a parsimonious explanation of how the process of late abstraction could enhance access to base analogs whose structural traits were not especially highlighted.

But even in those cases where participants are inclined to distill a more abstract representation of the sources, the sub-ceiling performance of interventions aimed at fostering an abstract encoding of the sources suggests that participants are unlikely to encode the relational structure of the source in ways that will neatly match the structure of the particular target they will have to solve. To illustrate, a proportion of participants might spontaneously encode the goal of capturing the fortress as an instance of "overpowering an object with a force," a representation that could later be useful to characterize the goal of destroying the tumor with rays in the radiation problem. But now consider that instead of confronting the radiation problem, participants had to figure out how to bring certain amount of water to a town while attending to the restriction that sending such an amount of water through either of the available canals alone would flood the surrounding lands. For this particular problem, it would have been more helpful to generalize the goal of capturing the fortress as a case of, say, "carrying a resource to a destination." There are numerous ways to encode exemplars (Hofstadter and FARG, 1995; Hofstadter & Sander, 2013; Markman & Ross, 2003), and people can hardly guess which one will best fit a target problem that they do not yet know. Even in cases like the above, an abstraction process could conceivably show a retrieval advantage over the raw processing of the target. Traditional models of analogical mapping have postulated several re-representation mechanisms aimed at revealing latent identities between initially non-identical elements (e.g., *minimal ascension*, Falkenhainer, Forbus, & Gentner, 1989; *coactivation over distributed representations*, Hummel & Holyoak, 1997, or *decomposition*, Gentner & Wolff, 2000). Employing a variant of minimal ascension, retrieval algorithms like ARCS (Thagard, Holyoak, Nelson, & Gochfeld, 1990) can access base analogs having similar—but not identical—concepts by means of probing LTM with memory cues that were derived from target elements through conceptual links such as superordination or part–whole relationships (e.g., the target concept *destroy* could activate a source containing the relation *capture* based on both being instances of *overpowering*). Given that connecting a superordinate concept to its instances requires a smaller number of links than connecting two particular instances (Gentner, 1989), interventions promoting late abstraction could in principle be capable of boosting the retrieval of base analogs whose structural encoding did not neatly match the structure of the target.

Yet another way in which an initial encoding could be considered suboptimal would consist in having stored a base situation under a poor or wrong principle, as compared to the more proper principle under which the target situation is currently being framed (Gentner et al., 2009). For instance, imagine that a professor wanted to illustrate the well-known *regression fallacy* (Tversky & Kahneman, 1974) with an example wherein the son of an extraordinary baseball player displayed skills that, although competitive, were not as outstanding as those of his father (according to the topic of the lesson, people's natural tendency to generate causal explanations for this kind of situations is ungranted, since such drops in performance could be more parsimoniously accounted for on purely statistical grounds). Now imagine that in an attempt to capitalize on her learners' prior knowledge, the professor asked them to recall further examples. Faced with this challenge, a sensible retrieval cue could involve a structure where the fact that a descendent of a high-performance sportsman did not display such level of performance gave rise to causal interpretations. By promoting a shift between a particular type of extreme values (e.g., striking baseball skills) and more general categories of outstanding performance (e.g., extraordinary levels of any kind of professional achievement), interventions aimed at eliciting target elaborations could conceivably enhance the retrieval of distant analogs that were indexed under a poor or wrong principle.

In sum, while Gentner et al. (2009) have modeled the beneficial effect of late abstraction for retrieving relationally identical sources whose surface content has not been sufficiently deemphasized, there are other types of suboptimal encoding that may complicate subsequent retrieval (Loewenstein, 2010), and which deserve further investigation. Despite the methodological difficulties mentioned above, we believe that the educational implications of the late abstraction principle justify any efforts directed to assess whether suboptimally encoded sources do in fact get retrieved, as well as the extent to which interventions directed to elicit an abstract encoding of the target can foster their retrieval.

## 7.8 Conclusions

Building incrementally on prior interventions credited with having promoted an abstract encoding of the base analogs, the studies reviewed in the present chapter have assessed whether applying such elaborations to the target analog could enhance the retrieval of distant sources, as predicted by the late abstraction principle. Most critically, our focus on self-accomplishable strategies was meant to overcome a serious applicability limitation of the target-comparison intervention: the need to provide analogizers with a specially tailored second analog for every new target situation they are to process. As these mental operations do not require the provision of additional information about the target, learners can potentially apply them across a wide variety of environments, both formal and informal.

## References

Belenky, D., & Schalk, L. (2014). The effects of idealized and grounded materials on learning, transfer, and interest: An organizing framework for categorizing external knowledge representations. *Educational Psychology Review, 26*(1), 27–50.

Bernardo, A. B. I. (2001). Analogical problem construction and transfer in mathematical problem solving. *Educational Psychology, 21*, 137–150.

Chen, Z., Mo, L., & Honomichl, R. (2004). Having the memory of an elephant: Long-term retrieval and use of analogues in problem solving. *Journal of Experimental Psychology: General, 133*, 415–433.

Duncker, K. (1945). On problem solving. *Psychological Monographs, 58* (5, Whole No. 270).

Dunbar, K. (2001). The analogical paradox: Why analogy is so easy in naturalistic settings, yet so difficult in the psychology laboratory? In D. Gentner, K. J. Holyoak, & B. Kokinov (Eds.), *The analogical mind: Perspectives from cognitive science* (pp. 313–334). Cambridge, MA: The MIT Press.

D'Angelo, V. & Trench, M. (2020). Enhancing distant analogical retrieval via generating abstract redescriptions of the target. Paper presented at the Conceptual Abstraction and Analogy in Natural and Artificial Systems Symposium of the Association for the Advancement of Artificial Intelligence, Washington, DC.

Falkenhainer, B. (1990). Analogical interpretation in context. In *Proceedings of the 12th Annual Conference of the Cognitive Science Society* (pp. 69–76). Hillsdale, NJ: Erlbaum.

Falkenhainer, B., Forbus, K. D., & Gentner, D. (1989). The structure-mapping engine: Algorithm and examples. *Artificial Intelligence, 41*, 1–63.

Finlayson, M., & Winston, P. (2006). *Analogical retrieval via intermediate features: The goldilocks hypothesis* (MIT CSAIL Technical Report No. MIT-CSAIL-TR-2006-071). hdl:1721.1/34635.

Gentner, D. (1989). The mechanisms of analogical transfer. In S. Vosniadou & A. Ortony (Eds.), *Similarity and analogical reasoning* (pp. 199–242). Cambridge, UK: Cambridge University Press.

Gentner, D., Loewenstein, J., & Thompson, L. (2004). Analogical encoding: Facilitating knowledge transfer and integration. In K. Forbus, D. Gentner, & T. Regier (Eds.), *Proceedings of the 26th Annual Conference of the Cognitive Science Society* (pp. 450–455). Mahwah, NJ: Erlbaum.

Gentner, D., Loewenstein, J., Thompson, L., & Forbus, K. (2009). Reviving inert knowledge: Analogical abstraction supports relational retrieval of past events. *Cognitive Science, 3*, 1343–1382.

Gentner, D., Rattermann, M. J., & Forbus, K. D. (1993). The roles of similarity in transfer: Separating retrievability from inferential soundness. *Cognitive Psychology, 25*, 431–467.

Gentner, D., & Wolff, P. (2000). Metaphor and knowledge change. In E. Dietrich & A. Markman (Eds.), *Cognitive dynamics: Conceptual change in humans and machines* (pp. 295–342). Mahwah, NJ: LEA.

Gick, M. L., & Holyoak, K. J. (1980). Analogical problem solving. *Cognitive Psychology, 12*, 306–355.

Hofstadter, D. R. (1985). *Metamagical Themas: Questing for the essence of mind and pattern.* London: Viking.

Hofstadter, D. R., & Sander, E. (2013). *Surfaces and essences: Analogy as the fuel and fire of thinking*. New York: Basic Books.

Hofstadter, D. R., & the Fluid Analogies Research Group. (1995). *Fluid concepts and creative analogies: Computer models of the fundamental mechanisms of thought*. New York: Basic Books.

Hummel, J. E., & Holyoak, K. J. (1997). Distributed representations of structure: A theory of analogical access and mapping. *Psychological Review, 104*, 427–466.

Keane, M. T. (1985). On drawing analogies when solving problems: A theory and test of solution generation in an analogical problem solving task. *British Journal of Psychology, 76*, 449–458.

Kurtz, K., & Loewenstein, J. (2007). Converging on a new role for analogy in problem solving and retrieval: When two problems are better than one. *Memory & Cognition, 35*, 334–341.

Lampert, M. (1986). Knowing, doing, and teaching multiplication. *Cognition and Instruction, 3*, 305–342.

Loewenstein, J. (2010). How one's hook is baited matters for catching an analogy. In B. H. Ross (Ed.), *The psychology of learning and motivation: Advances in research and theory* (Vol. 53, pp. 149–182). San Diego, CA: Elsevier.

Mandler, J. M., & Orlich, F. (1993). Analogical transfer: The roles of schema abstraction and awareness. *Bulletin of the Psychonomic Society, 5*, 485–487.

Markman, A. B., & Ross, B. H. (2003). Category use and category learning. *Psychological Bulletin, 129*, 592–613.

Medin, D. L., & Ross, B. H. (1989). The specific character of abstract thought: Categorization, problem-solving, and induction. In R. J. Sternberg (Ed.), *Advances in the psychology of human intelligence* (Vol. 5, pp. 189–223). Hillsdale, NJ: Erlbaum.

Minervino, R., Olguín, V., & Trench, M. (2017). Promoting interdomain analogical transfer: When creating a problem helps to solve a problem. *Memory & Cognition, 45*, 221–232. https://doi.org/10.3758/s13421-016-0655-2

Minervino, R., Trench, M., & Oberholzer, N. (2009). Concrete and imagined simulations of situation models enhance transfer of solutions to structurally different algebra word problems. In N. A. Taatgen & H. Van Rijn (Eds.), *Proceedings of the 31st Annual Conference of the Cognitive Science Society* (pp. 1394–1399). Austin, TX: Cognitive Science Society. MORAVIA 82.

Nathan, M. J., Kintsch, W., & Young, E. (1992). A theory of algebra word problem comprehension and its implications for the design of computer learning environments. *Cognition and Instruction, 9*, 329–389.

Rozenblit, L., & Keil, F. (2002). The misunderstood limits of folk science: An illusion of explanatory depth. *Cognitive Science, 26*(5), 521–562.

Rudnitsky, A., Etheredge, S., Freeman, S. J. M., & Gilbert, T. (1995). Learning to solve addition and subtraction word problems through a structure-plus-writing approach. *Journal for Research in Mathematics Education, 26*, 467–486.

Thagard, P., Holyoak, K., Nelson, G., & Gochfeld, D. (1990). Analog retrieval by constraint satisfaction. *Artificial Intelligence, 46*, 259–310.

Trench, M., Tavernini, M., & Goldstone, R. L. (2017). Promoting spontaneous analogical transfer by idealizing target representations. In G. Gunzelmann, A. Howes, T. Tenbrink, & E. Davelaar (Eds.), *Proceedings of the 39th Annual Conference of the Cognitive Science Society* (pp. 1206–1211). Austin, TX: Cognitive Science Society.

Tversky, A., & Kahneman, D. (1974). Judgment under uncertainty: Heuristics and biases. *Science, 185*, 1124–1131.

Ward, T. B., Smith, S. M., & Finke, R. A. (1999). Creative cognition. In R. J. Sternberg (Ed.), *Handbook of creativity* (pp. 189–212). Cambridge: Cambridge University Press.

# Chapter 8
# Epilogue: Unanswered Questions and Future Challenges in Creative Analogical Retrieval

Educators have been long preoccupied about students' difficulties to transfer their learning across thematic and contextual barriers, difficulties referred to as the "problem of inert knowledge". Given the centrality of analogical retrieval for flexible transfer of learning, realistic approaches to the problem of inert knowledge demand an accurate estimation of our ability to retrieve distant analogs from memory.

This ability lies at the intersection of the study of memory and reasoning functions, two areas that underwent rapid development during the outset of the information-processing approach. But despite their rapid development during the early years of the cognitive paradigm, theoretical considerations about the Quinean nature of the central systems' high-level cognitive functions (Fodor, 1983) rendered both areas suspect of lying beyond the reach of carefully controlled empirical investigations and attendant theory development. The body of behavioral and computational studies of which this monograph constitutes a rather incremental addition attests to the power of Cognitive Science to unveil the mysteries of high-level cognition. Despite the success of this interdisciplinary approach to analogical thinking, important questions remain unsettled.

Within the subprocesses of analogical thinking, the retrieval of source analogs from LTM is arguably the least understood, especially when it comes to explaining the retrieval of structural analogs devoid of surface similarity with the target. Not only are there different theoretical models about the unobservable mental operations on which it is based, but also fundamental disagreements about our basic ability (or lack thereof) to retrieve distant analogs from memory.

At one side, traditional experiments using a two-phase transfer paradigm consistently documented the improbability of retrieving analogous cases in the absence of at least some minimal amount of surface similarities. On the other side, a rapidly growing body of naturalistic studies reveals a frequent use of far analogies by experts and novices, a result that calls into question both the validity of

 
M. Trench, R. A. Minervino, *Distant Connections: The Memory Basis of Creative Analogy*, SpringerBriefs in Psychology,
https://doi.org/10.1007/978-3-030-52545-3_8

experimental studies and the adequacy of the computer models that were engineered to mimic such patterns of results.

After discussing the strengths and weaknesses of the experimental and the naturalistic approaches, we presented our own attempt to decide between these accounts by means of a hybrid paradigm that retained both the ecological validity of naturalistic studies and the methodological control of experimental procedures. In line with traditional experimental results, the application of this hybrid paradigm to three different kinds of naturally acquired sources (popular movies, highly publicized political affairs, and autobiographical episodes) converged in demonstrating that surface similarity governs analogical retrieval even when ecologically valid tasks and materials are involved. Given that our retrieval mechanisms do not seem to distinguish the natural from the artificial, the inspection of natural settings should no longer be regarded as a privileged environment wherein to search for the mysterious ingredients that allow for flexible transfer.

In light of the limitations of our memory systems for accessing far analogs, the second half of the present monograph began by reviewing early attempts to make distant connections more likely. Based on the intuition that disembedding the initial learning from their context-specific information should render it more retrievable during the future processing of superficially dissimilar targets, several interventions were aimed at enforcing a more abstract encoding of the base analogs. Despite the success of a handful of these strategies within the vacuum chamber of the psychological laboratory, the question remains as to which of these interventions would survive the more noisy environments that characterize instructional settings. But even if source abstraction proved amenable to realistic instructional environments, interventions following this approach have little to offer to someone needing to retrieve base situations whose encoding conditions had not emphasized their abstract features.

The last two chapters review recent attempts to help participants retrieve this particular kind of stored situations. Albeit commonsensical, one of the avenues we began to explore concerned determining whether the deliberate intention to search for analogous cases increases the probabilities of retrieving distant analogs from LTM. Results from a series of experiments carried out in our laboratory suggest that while the deliberate search can increase access to distant sources during argumentation and hypothesis-generation activities, it does not prove effective during problem-solving. We conjectured that participants react to impasses in obtaining satisfactory solutions by consciously searching for analogous cases, but the nature of the representations that typically serve as targets of problem-solving makes it hard to locate distant sources despite this metacognitive predisposition. Hence, boosting access to distant sources in the service of problem-solving requires interventions somewhat more sophisticated.

A promising line of intervention builds upon the late abstraction principle postulated by Dedre Gentner and colleagues. In an apparent contradiction with the widely accepted *encoding specificity hypothesis* (Tulving & Thompson, 1973), it states that the benefits of generating abstract representations should also apply to elaborations of the target analog at retrieval time. Since recently, the only evidence for the psychological reality of late abstraction came from Kurtz and Loewenstein (2007) and Gentner, Loewenstein, Thompson, and Forbus (2009), who elicited a retrieval advantage by presenting the target analog together with another unsolved problem with similar structure, and having participants compare them. Enthused by the potential of late abstraction to overcome the applicability limitations of interventions based on initial encoding, we developed a series of interventions designed to help participants capitalize on the late abstraction principle, but without depending on the external provision of target-specific information. While the task of comparing the target to a random non-isomorphic problem proved unsuccessful, activities like inventing a new problem with a similar structure, constructing idealized perceptual representations of the target problem, or producing functional definitions of the problem's central objects increased participants' ability to transfer the base solutions to the target problem.

The fact that analogical retrieval could be boosted either via conceptual or perceptual abstractions of the target suggests that, although essentially correct, current computational simulations of the late abstraction principle may still be missing a perceptual dimension of how base and targets are being represented. Along with other studies documenting the role of perceptually represented structure in retrieval (e.g., Catrambone, Craig, & Nersessian, 2006; Clement, 2008; Day & Goldstone, 2011; Grant & Spivey, 2003; Pedone, Hummel, & Holyoak, 2001), the present results signal the need of a broader, unifying account of analogical retrieval and transfer.

Despite the excitement about the prospects of target elaborations for facilitating access to information that would have otherwise remained dormant, several caveats are in order. On the one hand, advocates of the well-established encoding specificity hypothesis are often skeptical about the potential of late abstraction for retrieving suboptimally encoded sources. As the retrieval rates elicited by target abstraction strategies are typically far from ceiling, it could be the case that their net improvements over their baseline conditions reduce to a minority of cases wherein the base analog has been spontaneously abstracted by participants. Hence, while it is undeniable that target abstraction augmented interdomain retrieval, the claim that it boosted access to suboptimally encoded sources might be ungranted. Even though we outlined a handful of ways by which non-abstracted sources could still be retrieved in response to the processing of distant analogs whose structural features had been highlighted (see Chap. 7), a definitive proof that target abstraction aids access to non-abstracted sources awaits further research. This kind of demonstration may not be around the corner, since it would require an independent assessment of how source situations were encoded in LTM or, more difficult yet, a demonstration that an abstract encoding had *not* taken place.

Another limitation of current behavioral evidence for the late abstraction principle concerns the almost exclusive deployment of the military and the tumor problems, "the fruit fly" of analogical retrieval. While employing the military and the tumor problem allows for a rather straightforward incorporation of a new finding within the context of prior studies, there is always the risk that such findings might fail to generalize.

Yet another self-imposed limitation in the service of clear-cut empirical results concerns the almost exclusive deployment of randomized samples of undergraduate participants. The assessment of individual differences in studies that would still make perfect sense without them is sometimes interpreted as providing a sort of life saver for the eventual absence of main effects. However, we believe that there is a lot yet to be learned from the study of how individual differences might affect analogical transfer. As an example, Kubricht, Lu, and Holyoak (2017) obtained that the effectiveness of coupling verbal descriptions of the military story with static diagrams or visual animations was modulated by participants' comprehension of the military story, which was in turn influenced by participants scores in Raven's Progressive Matrices (RPM) test. One relatively overlooked result from their first experiment concerns the contrast between the retrieval rates of high RPM vs. low RPM participants in the baseline condition, wherein no visual supports were provided (73% vs 40%). While it is certainly possible that this difference stems from a more complete (but still concrete) understanding of the base situation, we speculate that high RPM participants might have encoded the base situation at a higher level of abstraction, thus rendering it more retrievable despite concrete-level mismatches with the target. After all, while there is some debate about how narrow or broad a definition of intelligence should be, the ability to think abstractly may still represent the single trait most unanimously associated with intelligence.[1] Quite recently, Navarrete (2017) found evidence that the probability of spontaneously transferring the nature of a visual animation to a contextually separated analogous situation was correlated with higher scores on the Cognitive Reflection Test (Frederick, 2005), an indicator of System 2's propensity to supervise intuitive solutions provided by System 1. Hence, it would seem that successful encoders would benefit from both a natural ability for abstraction—as measured by RPM—as well as from a metacognitive predisposition to reflect on the input they receive.

## 8.1 Is There a Basic Level of Abstraction?

As stated at the end of Chap. 7, we are currently developing instructional materials aimed at helping university-level participants hone in on an appropriate level of abstraction for a base or target situation,[2] which we tentatively defined as being the

[1] Other individual traits allegedly facilitating distant analogical transfer are field-independence (Antonietti & Gioletta, 1995) and divergent associative thinking, as measured by the remote associates task (Cushen & Wiley, 2018). However, some methodological decisions make these studies look like a test of mapping rather than of retrieval: While the former included a hint to consider the source stories, the second employed a base analog that was thematically related to the tumor problem (the light-bulb story).

[2] In constructing our examples, the gradation of generality involved both objects and relations. Even though this criterion renders the task more challenging than simply abstracting entities and contextual features—see, e.g., MAC/FAC simulations—we do this under the assumption that in analogies of realistic complexity, the core relational predicates are seldom identical across analogs.

most general formulation that still retains all critical features of the original situation. The first part of the training confronts participants with commented examples wherein given problems or situations were used as a basis to derive abstract schemas of different degrees of generality. For example, consider a hypothetical situation in which a professor felt he was capable of directing the teacher training program where he was working, but faced the problem that the director of the program had great publications and was relatively young, such that he did not foresee realistic ways of getting the leading position. Based on this original situation, the following formulations would represent different degrees of generality:

**Minimal Generalization** A situation wherein a scholar feels capable of directing the educational project where he is working, but faces the problem that his superior is productive and relatively young, such that he does not foresee realistic ways of ascending.

**Intermediate Generalization** A situation wherein a person or institution feels capable of occupying a given position in the organization to which it belongs, but faces the problem that the person or entity holding the desired position has greater antecedents, such that he or it does not foresee realistic ways of obtaining it.

**Excessive Abstraction** A situation wherein a person or institution feels unsatisfied with his position in the organization to which he belongs and wishes to hold another position.

Using a "fading" schedule, subsequent sections ask participants to order shuffled arrays of an original situation + three reformulations in terms of their degree of generality and finally to classify further pairs of a situation + a single reformulation as involving minimal, intermediate, or extreme degrees of generality. Pilot tests are revealing that it is quite hard for undergraduate participants to master this ability by means of a reasonably concise training scheme (ideally, a single session) in the lack of personalized feedback. Despite having been exposed to many commented examples of how an initial situation could be generalized to different degrees, participants are both intra- and interpersonally inconsistent with regard to their preferred level of generality.

But how to converge on an appropriate level of abstraction without knowing how distant a stored (or future) analog might be? For clarity, let us begin by considering the problem of how to encode a current event for future use. Sticking to the above-mentioned "ambitious professor" scenario, let us assume that the story features a creative solution that the encoder renders potentially exportable: the jealous professor shifted from circulating his own CV to publicizing that of his superior, with the result that his boss was offered a better job and moved away. Conservative generalizations will be more powerful than relatively more general ones for enabling the retrieval of the original situation during the future processing of a close analog (e.g., a researcher's intention to be the principal investigator of the laboratory where she is working, which is currently led by a more experienced researcher). On the flip side, these narrow generalizations will be less effective during the processing of novel situations that are rather distant (e.g., Italy's frustration for not being able to occupy the leading role in UEFA, due to Germany having even more impressive credentials). But if Kubricht et al.'s results are to be taken as evidence that more intelligent participants are more effective than their low-RPM peers in calibrating the level of generality at which encountered situations should be encoded, then there should be some principled way of determining, for any given situation, which level of abstraction would be more convenient, even if just probabilistically. The ability to hone in on the most transfer-effective generalization will probably demand both world knowledge and imagination: Quite recursively, reasoner's attempts to derive a sense of how general a formulation ought to be might involve evoking or else envisioning potential analogs in other domains and adjusting abstraction so as to encompass them without slipping into overly abstract schemas. If the original event gets mentally revisited in later times, there is the chance that it can elicit alternative generalizations, further increasing its future retrievability.

While different in appearance, the case of generalizing a target situation for accessing stored situations is similar in essence, with the main difference being chiefly one of specificity: if the target situation is indeed pressing, then it is unlikely that it will belong to a category of events for which suitable analogs abound. In these cases, deciding between alternative degrees of generality might reduce to imagining potential analogs and/or squeezing one's world knowledge and intuition so as to assess the promise of alternative domains wherein the central aspects of the target situation could potentially apply.

To complicate matters more, results from simulation experiments suggest that the *size* of target representations can also influence analogical access, with intermediate-sized representations being more powerful than either small or large ones for supporting expert-like analogical retrieval (the *Goldilocks hypothesis*, Finlayson & Winston, 2006). This computational consideration could bear potential implications for the prospects of abstraction, since we had informally observed that the size of the schemas that can be derived from an original situation varies non-linearly with their level of abstraction: As shown by the ambitious professor example, minimal abstractions are about the same size of the original situation, they get larger at intermediate levels of generality and finally drop steeply in size when they get too vapid. Future behavioral studies should address exactly which combinations between representational size and level of generality maximize the probability of retrieving distant analogs from memory, as well as whether (and eventually how) the esoteric ability to optimally calibrate these variables could possibly be trained.

## 8.2 Beyond Strict Analogical Retrieval

In our discussion of the factors that facilitate access to distant sources, we have committed to a rather strict definition of analogical retrieval: the activation of a base analog from LTM during the processing of a specific target analog in WM. While it is certainly plausible that creative individuals identify far analogs in LTM by means of generating more general, decontextualized representations of base and target situations, the generation of distant analogies can also be aided by several mechanisms that do not conform to such a conservative definition.

A particularly interesting way of generating distant analogies is the reuse of *conceptual metaphors* (Lakoff & Johnson, 1980), that is, culturally shared analogies between an abstract target domain (e.g., the domain of *understanding*) and a relatively more concrete, perceptually grounded domain (e.g., the domain of *vision*). In a recent experimental study on the use of analogies during argumentation, Olguín, Trench, and Minervino (2017) found that even though participants were explicitly asked to base their analogies on autobiographical episodes, many of the distant analogies produced by participants were in fact *extensions* of conceptual metaphors. Within a conceptual metaphor, only some aspects of the base domain are projected onto the target, while other aspects tend to be left aside (Kovecses, 2002; Lakoff, 1990; Turner, 1996). One heuristic used by experts for generating novel analogies consists in extending the standard mapping between the base and the target domain. As conceptual metaphors connect distant domains (e.g., *discussion* is thought as *war*, Lakoff & Johnson, 1980), extending conceptual metaphors results in novel instances of a culturally shared distant analogy. These extensions give rise to products like metaphorical expressions (Lakoff & Turner, 1989), idioms (Gibbs, 1994), stories (Turner, 1996), or images (Forceville, 2006), which can be employed in the service of literature writing, teaching, humor, advertising, or argumentation. As an example, in the A PROJECT IS A JOURNEY conceptual metaphor, some of the most frequently mapped aspects of the base domain of traveling are the vehicle, the destination, the obstacles faced during the trip, or the speed of the vehicle. Subtler aspects related to the base domain (e.g., the landscapes we see while traveling) are not usually transferred to the target. Just to exemplify with one of the naturalistic studies commented in Chap. 3, in Blanchette and Dunbar (2001) some of the analogies that appeared in the press prior to the 1996 referendum on the independence of Quebec were created by extending conceptual metaphors. For instance, one of the anti-emancipation analogies compared emancipation to the act of abandoning an ocean liner to board on a lifeboat in the middle of a storm, an analogy that involves extending the conceptual metaphor A PROJECT IS A JOURNEY (the change of vehicle in response to traveling difficulties is not usually employed as a base domain when thinking about projects).

In an earlier study on the use of metaphors for creative titling (Trench & Minervino, 2015), we assessed whether novices could be trained in identifying culturally shared metaphors for a given topic as well as in reinstantiating and extending conceptual metaphors, two heuristics formerly observed among expert writers. After the training sessions, participants of the experimental group were handed three target passages, each one dealing with a different topic, and were encouraged to produce creative titles through the learned heuristics. Their titles were compared to those of a control group that had received a basic instruction about the concept of metaphor, as well as the instruction to generate creative metaphorical titles, but no training on the particular heuristics. In line with cognitive accounts of creativity as involving a perfected deployment of otherwise ordinary mental processes (e.g.,

Weisberg, 2006), results showed not only that the trained group produced a greater proportion of metaphoric extensions, but also that their extensions were rated more creative than those that were spontaneously produced in the untrained condition. Hence, analogies arising from a strategic reuse of distant connections previously established by others can be effective as well as creative.

When attempts to generate analogies by probing either one's own memory or that of the culture do not yield satisfactory outcomes, suitable analogs can eventually be provided by the environment. During *creative worrying* (Browne & Cruse, 1988; Olton, 1979), the reasoner occasionally interleaves daily activities with lapses of time during which she revisits the problem that remained unsolved. During those lapses, the social or physical environment can serendipitously present the reasoner with a relevant analog, leading to the establishment of a useful analogy. Just for the sake of illustration, the philosopher Cornelius Castoriadis has famously compared the irresponsible devastation of natural resources by capitalist societies with a central episode from Hansel and Gretel's tale, where the starving protagonists devoured the chocolate walls of their house without realizing that they were compromising their shelter. In wondering how he came up with such an analogy, one could picture him in his home environment, recurrently switching attention back and forth between his domestic assignments and his political elucubrations about irresponsible consumerism in modern societies. It is easy to see how the reading of Hansel and Gretel to his grandchildren would have elicited the coactivation of both analogs in working memory, leading to an easy mapping between how capitalist societies exhaust natural resources and how Hansel and Gretel ate the chocolate walls of their shelter.

Most research on analogical retrieval has tended to involve sources and targets maintaining high degrees of structural isomorphism. But real-world analogies break down at some point, and the reasoner is faced with deciding whether to drop the analogy altogether, or else to get involved in a creative exploration and eventual re-representation of the compared domains. In a celebrated observational study, John Clement had ten expert physicists decide which of two springs would stretch more under the same force. Both springs were made of the same kind of wire, with the same number of coils and the same length, with the only difference being that the coils of the former were twice as wide as those of the latter. Clement observed that 18 out of the 31 significant analogies produced by participants were "generated via a transformation" of the original situation, thereby changing one or more features that were previously assumed to be fixed (Clement, 1988, 2008). For example, one subject considered uncoiling the original springs so as to transform them into a straight horizontal wire, with the weight on its end, whereas another subject envisioned the case of further bending a horizontal wire so as to produce square-shaped coils and used these hypothetical cases as a source for correctly predicting that the wider spring would stretch more.

But imposing a transformation on the target so as to convert it into something else is a rather conservative option, possibly selected into response to the stringent task of predicting the behavior of a system whose internal rules of functioning one does not fully comprehend. As speculated in Chap. 6, less constrained pragmatics such as persuasive argumentation typically involve target situations that, albeit novel for the recipient of the analogy, are reasonably "complete" and well-known by the analogizer. As an example, the situation of adopting a zero-deficit strategy lends well to distilling an abstract schema such as "when something that grows geometrically is not controlled in time, at a later time it can grow so rapidly that attempts to control it can potentially fail." In cases like this, suitable source analogs can be generated simply by identifying negative elements which grow sort of geometrically, and embedding them within a narrative wherein delayed attempts to control them ended up being unfruitful.

As these cases illustrate, creative analogizing can sometimes circumvent the need of retrieving completely isomorphic representations from memory, or of opportunistically assimilating analogous cases serendipitously presented by the environment. In combination with the surface bias of our memory systems, these collateral mechanisms could potentially explain why a journalist would probably fail to retrieve Gick and Holyoak's (1980) military story during a typical experimental study of analogical retrieval but yet succeed in coming up with a couple of smart analogies during the writing of a newspaper article. Future studies involving non-expert populations should delve deeper into how analogy fabrication adapts to the constraints imposed by cognitive activities as varied as solving a problem, generating explanatory hypotheses, predicting the probability of hypothetical events, or communicating ideas to others, as well as into the mental operations that are recruited in order to satisfy these constraints.

The contribution of memory functions to creative analogizing would intuitively seem to be more strongly associated with the traditional definition of analogical retrieval than to opportunistic assimilation or even invention. But long-term memory is widely considered to be the most transversal of cognitive functions, inevitably subserving all other functions and especially high-level cognition. Understanding and eventually modeling how memory and executive functions jointly support these alternative routes to analogical transfer represents a major challenge for Cognitive Science and Artificial Intelligence. Despite more than three decades of analogy research, it would seem that the quest to defy Fodor's somber premonition about the decipherability of high-level cognition is still in its beginnings.

## References

Antonietti, A., & Gioletta, M. A. (1995). Individual differences in analogical problem solving. *Personality and Individual Differences, 18*, 611–619.

Blanchette, I., & Dunbar, K. (2001). Analogy use in naturalistic settings: The influence of audience, emotion, and goals. *Memory & Cognition, 29*, 730–735.

Browne, B. A., & Cruse, D. F. (1988). The incubation effect: Illusion or illumination? *Human Performance, 1*, 177–185.

Catrambone, R., Craig, D. L., & Nersessian, N. J. (2006). The role of perceptually represented structure in analogical problem solving. *Memory & Cognition, 34*, 1126–1132.

Clement, J. (1988). Observed methods for generating analogies in scientific problem solving. *Cognitive Science, 12*, 563–586.

Clement, J. (2008). *Creative model construction in scientists and students*. New York: Springer.

Cushen, P. J., & Wiley, J. (2018). Both attentional control and the ability to make remote associations aid spontaneous analogical transfer. *Memory & Cognition, 46*, 1398–1412.

Day, S., & Goldstone, R. L. (2011). Analogical transfer from a simulated physical system. *Journal of Experimental Psychology: Learning, Memory, and Cognition, 37*, 551–567.

Finlayson, M., & Winston, P. (2006). Analogical retrieval via intermediate features: The Goldilocks hypothesis. MIT CSAIL Technical Report No. MIT-CSAIL-TR-2006-071. hdl:1721.1/34635.

Fodor, J. (1983). *The modularity of mind*. Cambridge, MA: MIT Press.

Forceville, C. (2006). Non-verbal and multimodal metaphor in a cognitivist framework: Agendas for research. In G. Kristiansen, M. Achard, R. Dirven, & F. Ruiz de Mendoza (Eds.), *Cognitive linguistics: Current applications and future perspectives* (pp. 379–402). New York: De Gruyter Mouton.

Frederick, S. (2005). Cognitive reflection and decision making. *Journal of Economic Perspectives, 19*, 25–42.

Gentner, D., Loewenstein, J., Thompson, L., & Forbus, K. (2009). Reviving inert knowledge: Analogical abstraction supports relational retrieval of past events. *Cognitive Science, 3*, 1343–1382.

Gibbs Jr., R. W. (1994). *The poetics of the mind: Figurative thought, language and understanding*. Cambridge: Cambridge University Press.

Gick, M. L., & Holyoak, K. J. (1980). Analogical problem solving. *Cognitive Psychology, 12*, 306–355.

Grant, E. R., & Spivey, M. (2003). Eye movements and problem solving: Guiding attention guides thought. *Psychological Science, 14*(5), 462–466.

Kovecses, Z. (2002). *Metaphor: A practical introduction*. Oxford, NY: Oxford University Press.

Kubricht, J. R., Lu, H., & Holyoak, K. J. (2017). Individual differences in spontaneous analogical transfer. *Memory & Cognition, 45*(4), 576–588.

Kurtz, K., & Loewenstein, J. (2007). Converging on a new role for analogy in problem solving and retrieval: When two problems are better than one. *Memory & Cognition, 35*, 334–341.

Lakoff, G. (1990). The invariance hypothesis: Is abstract reason based on image-schemas? *Cognitive Linguistics, 1*, 39–75.

Lakoff, G., & Johnson, M. (1980). *Metaphors we live by*. Chicago: Chicago University Press.

Lakoff, G., & Turner, M. (1989). *More than cool reason: A field guide to poetic metaphor*. Chicago: University of Chicago Press.

Navarrete, J. A. (2017). Individual differences in transfer mediated by conceptual priming. In G. Gunzelmann, A. Howes, T. Tenbrink, & E. Davelaar (Eds.), *Proceedings of the 39th Annual Conference of the Cognitive Science Society* (pp. 841–846). Austin, TX: Cognitive Science Society.

Olguín, V., Trench, M., & Minervino, R. (2017). Attending to individual recipients' knowledge when generating persuasive analogies. *Journal of Cognitive Psychology, 29*, 755–768.

Olton, R. M. (1979). Experimental studies of incubation: Searching for the elusive. *Journal of Creative Behavior, 13*, 9–22.

Pedone, R., Hummel, J. E., & Holyoak, K. J. (2001). The use of diagrams in analogical problem solving. *Memory and Cognition, 29*, 214–221.

Trench, M., & Minervino, R. (2015). Creativity training from a continuist perspective: Reviving dormant analogies to generate novel metaphorical expressions. *Creativity Research Journal, 27*, 188–197.

Tulving, E., & Thompson, D. M. (1973). Encoding specificity and retrieval processes in episodic memory. *Psychological Review, 80*, 352–373.

Turner, M. (1996). *The literary mind: The origins of thought and language*. Oxford: Oxford University Press.

Weisberg, R. W. (2006). *Creativity: Understanding innovation in problem solving, science, invention, and the arts*. Hoboken, NJ: Wiley.

# Index





M. Trench, R. A. Minervino, *Distant Connections: The Memory Basis of Creative Analogy*, SpringerBriefs in Psychology,
https://doi.org/10.1007/978-3-030-52545-3

The manufacturer's authorised representative in the EU is Springer Nature Customer Service Centre GmbH, Europaplatz 3, 69115 Heidelberg, Germany. If you have any concerns regarding our products, please contact ProductSafety@springernature.com

Printed and bound by CPI Group (UK) Ltd, Croydon, CR0 4YY
22/09/2026
02221064-0017